THE SUPERPOWER
TRANSFORMATION

THE SUPERPOWER TRANSFORMATION

Making Australia's zero-carbon future

EDITED BY

ROSS GARNAUT

LA TROBE
UNIVERSITY PRESS

IN CONJUNCTION WITH BLACK INC.

Published by La Trobe University Press in conjunction with Black Inc.
22–24 Northumberland Street
Collingwood VIC 3066, Australia
enquiries@blackincbooks.com
www.blackincbooks.com
www.latrobeuniversitypress.com.au

La Trobe University plays an integral role in Australia's public intellectual life, and is
recognised globally for its research excellence and commitment to ideas and debate.
La Trobe University Press publishes books of high intellectual quality, aimed at general
readers. Titles range across the humanities and sciences, and are written by distinguished
and innovative scholars. La Trobe University Press books are produced in conjunction
with Black Inc., an independent Australian publishing house. The members of the LTUP
Editorial Board are Vice-Chancellor's Fellows Emeritus Professor Robert Manne and
Dr Elizabeth Finkel, and Morry Schwartz and Chris Feik of Black Inc.

9781760643843 (paperback)
9781743822524 (ebook)

A catalogue record for this
book is available from the
National Library of Australia

Cover design by Akiko Chan
Cover image: Cavan Images / Shutterstock
Text design by Dennis Grauel and Typography Studio
Typesetting by Typography Studio

Printed in Australia by McPherson's Printing Group

CONTENTS

CONTENTS

CONTENTS

INTRODUCTION

Ross Garnaut

Several historic events have placed Australia in a strong position to grasp the opportunity described three years ago in my book *Super-power: Australia's Low-Carbon Opportunity*. This book illuminates the opportunity and how Australia can make the most of it. Success will shift Australia from a laggard in the world's efforts to deal with the climate change threat to a leader. It will support the restoration of our own prosperity after a decade of stagnation of living standards.

The election of 21 May 2022 brought majorities for effective action on climate change into the House of Representatives and the Senate. The new prime minister, Anthony Albanese, said in his acceptance speech that his government would make Australia a renewable energy Superpower. A greatly enlarged crossbench comprises people of exceptional ability committed to supporting the Australian climate and energy transition.

The Russian invasion of Ukraine in February 2022 cuts across every contemporary issue. The disruption of Russian coal and gas exports created scarcity in Europe and lifted global prices for fossil carbon several-fold. European governments hesitated, and then committed decisively to accelerating movement to the new economy. Australian renewable resources quickly came into focus as a reliable source of the green hydrogen and zero-carbon goods that could not be provided in adequate volumes at home. The high global prices for coal and gas found their way into Australian domestic prices and threatened to disrupt our own prosperity. We, too, face a choice between doubling down on the old and accelerating movement to the new.

Superpower Transformation brings together knowledge of different aspects of the Superpower opportunity. It shows that the immense demands of the new energy systems and industries can provide all of the large increase in investment in export industries in Australia that is necessary to achieve sustainable full employment with rising living standards. It shows that Australia can supply zero-emissions goods and services that directly reduce global emissions by around 8 per cent – much more than Europe (including the UK) achieving zero net emissions, or more than twice Japan, or more than India doing so. Australia's success would materially improve the world's chances of avoiding increases in temperatures so large that they threaten Australian prosperity, global development and the political order upon which we all depend.

In an overview essay in Chapter 1, I describe the Australian journey on climate and energy policy over the past one and a half decades, the immense obstacles that we have placed in our own path, and the new chance to move past the blockages. The historical legacy is heavy, and not easily pushed aside. But the new circumstances in our community and parliament and the dimension of the opportunity make it possible to do so. The new government's policies allow us to make a strong start. The chapter outlines the policies that will enable us to finish.

Australia has been locked in climate wars over the past decade. Recognition of economic opportunity enables the large majority of Australians to join a productive peace. Chapter 1 and the book as a whole describe the peace dividend.

Until now, we have engaged in our internal discussions as if playing our full part in the global effort to hold the damage from climate change within reasonable limits involved a sacrifice of our own welfare. Somehow we would benefit if we did less, or delayed until later anything we do. Chapter 1 shows that there are net economic benefits in moving to zero emissions on early timetables – on electricity by the mid-2030s and in the whole economy a decade after that. There are adjustment stresses that must be absorbed at some time. The earlier they are taken, the sooner Australians can enjoy insulation of domestic energy prices from the vagaries of global fossil energy markets, and the benefits from investment in and export of the low-emissions products of the future.

At the Glasgow meeting, support coalesced around the objective of holding human-induced temperature increases to 1.5°C. In Chapter 2, Meinshausen et al. define authoritatively the extent and timing of emissions reductions necessary to achieve this goal – for the world to hold temperature to 1.5°C. They also discuss the developed world's and Australia's fair share of the global effort. The world as a whole will need to achieve net zero by 2050. It is accepted that equity requires developed countries to move before developing. The arithmetic of atmospheric physics requires developed countries to reach net zero by 2045. Only a few developed countries have yet committed to that goal, with Germany and some other European countries leading the way. That would allow India (now committed to net zero by 2070) and other developing countries to take until the mid-2060s. Russia and China advised the Glasgow meeting that they would achieve net zero by 2060. The global arithmetic adds up to net zero if they bring forward their own objective to the mid-2050s.

Dylan McConnell in Chapter 4 discusses how the electricity market has developed in Australia over the past decade. Supportive policies, notably the Renewable Energy Target (including its small-scale component), have led to strong growth in renewable energy supply. This has been decelerating in recent years, and in the absence of new policy will fall well short of delivering enough solar and wind generation to meet the government's goals for expansion of renewable energy and reduction of emissions.

In Chapter 5, Frank Jotzo demonstrates how the conversion of renewable energy into hydrogen ('green hydrogen') in Australia can play a major role in the decarbonisation of this country and the world. Not all hydrogen is the same. Alternative processes using gas or coal as a feedstock ('blue', 'grey', 'black' and 'brown' hydrogen) have high emissions, which are unlikely to be removed by carbon capture and storage. External benefits from pioneering companies' innovation in commercial development of hydrogen warrant public fiscal support for early projects. The potential scale of developments is immense. It is important to think now about environmental amenity and fair treatment of Indigenous and other local communities if we are to avoid the problems of the old resources industries.

The largest of the hydrogen opportunities for Australia is in converting iron ore into metal. Ligang Song in Chapter 6 and Mike Sandiford in Chapter 3 note that conversion of Australian iron ore into metal for export is economically feasible at the current high prices for hydrogen's established alternative as a reductant for iron: metallurgical coal. Premiums are emerging for zero-emissions iron and steel, which enhances commercial returns. Traditional steel-making processes generate around 7 per cent of global greenhouse gas emissions. The most likely path to eliminating most of these at low cost is through direct reduction using hydrogen. For many countries, the processing of iron ore into metal will be much less costly in Australia than at home. If other countries undertake early-stage processing in Australia, this would reduce global greenhouse gas emissions by several times as much as Australia reducing its own emissions to zero.

China produces and uses more than half the world's iron and steel, most of it from Australian iron ore. In Chapter 6, Ligang Song discusses the large role that Sino–Australian cooperation could play in Chinese and global decarbonisation. China's steel industry is responsible for around 4 per cent of total global emissions. Reduction of ore into metal will be achieved most efficiently through use of hydrogen. This will be a large challenge in China, given competing demands for zero-emissions electricity. Shifting Chinese imports from Australian ore to Australian metal would lower the costs of decarbonisation in China and add immensely to Australian development. The changes in attitudes to inter-dependence would be similar in kind and smaller in scope than with the growth in Chinese reliance on imported iron ore in the late twentieth and early twenty-first centuries. Realisation of this great potential for Chinese and Australian decarbonisation and Australian development would require the restoration of cooperative and productive trade relations.

The global transformation of energy requires immense supply of minerals that currently are not available in large quantities. These have come to be called 'critical minerals'. Following Sandiford in Chapter 3, we prefer to call them 'energy transition minerals'. Demand for them is estimated to grow seven-fold by 2030 and much more after that. Australia has abundant resources of critical minerals. Many of them require a great quantity of energy – in the zero-emissions economy, renewable energy – for processing. Sandiford discusses the Australian opportunity to become the world's major supplier for a range of energy transition minerals in raw and processed form to international markets. These developments draw on old Australian skills, infrastructure and mining and industrial capacities. Effective development of Australian capacities in these areas can ease what otherwise would be tight bottlenecks slowing global decarbonisation. At the same time, it would underwrite Australian development, much in regional areas challenged by decline in fossil carbon production and use.

Alongside Australia's rich solar and wind resources, the potential to sequester carbon in the landscape and grow biomass is a source of comparative advantage in international trade. In Chapter 7, Isabelle Grant takes a practical look at the possibilities. Australia has the developed world's largest per-capita endowment of land suitable for growing plants for biomass and absorbing carbon in the landscape. Research, development and commercialisation are in early stages. The chapter provides examples of valuable use of several native Australian acacias and eucalypts, adapted to Australia's variable rainfall and high temperatures. It also looks at the use of one exotic species, agave, adapted for hot and dry climates in Mexico. Use of a small proportion of available land would yield large amounts of carbon for permanent storage out of the atmosphere or as inputs for zero-emissions chemical industries and thermal energy for transport or peaking electricity generation.

Development of Australia's Superpower opportunity will have its greatest positive impact on employment and incomes in rural and provincial Australia. Old industrial areas will have new leases of life. New industrial precincts will emerge in rural areas. In Chapter 8, Susannah Powell presents a case study: the Barcaldine Renewable Energy Zone industrial precinct in central west Queensland, in the shire that covers the Galilee Basin and controversial coalmines. Planning has identified an opportunity roughly to double employment in an old town built around the wool industry and that has experienced many decades of stagnation and decline. Rich solar and wind resources and availability of waste biomass and opportunities to supplement that with new growth have the potential to support zero-emissions production of urea (Australia's and one of the world's first zero-emissions nitrogenous fertiliser plants), large-scale protected horticulture, a range of industries applying pyrolysis and distillation technologies to transform local wastes and cultivated biomass into industrial inputs, a manufacturer of stock feed for cattle and sheep, and processing of local mineral resources. The case study reveals an immense coordination task.

Success as the Superpower of the low-carbon world economy can make a large contribution to restoring Australia after the stagnation of the Dog Days and the stress of the pandemic recession. It will help to restore productive relations in our Southwest Pacific neighbourhood and lift our standing more broadly in the international community. And it will contribute significantly to humanity's chances of combating a threat to peace and prosperity.

THE BRIDGE TO THE SUPERPOWER

Ross Garnaut

A new political landscape

There is a place of calm and beauty on the Alice River, half an hour by bicycle from Barcaldine on the road to the Barcoo. Solid park benches are fixed in the shade of the stately river gums. Trunks, branches and leaves are set in perfect detail against a cobalt-blue sky reflected in a riverbed pool. That's where I prepared the lectures in the winter of 2020 that became the book *Reset: Restoring Australia after the Pandemic Recession*. If I started early, a wallaby and then an emu would rustle past for a morning drink, startling to flight the teal ducks. A wedgetail floated above, keeping a silent eye on opportunity. A couple of pelicans would glide towards me, to greet and observe as I opened the laptop. Here and there, the water would be broken by the yellowbellies taking their chances with a morning frolic. The pool was slowly shrinking day by day, exposing more of the fish. More and more tangled logs and branches and cracks appeared in the mud around the receding water's edge.

But in 2022, a sequence of unusual rain events changed inland Queensland. The Alice and all the channels west and south breached their banks and pushed the muck out of the way. The vast plains of Mitchell grass turned emerald-green on the grey clay soils. New leaves brightened the gidgee, bloodwood, whitegum and boree in the sandy country. When the rain stopped, the overflow receded to a thousand veins of mercury glistening in the sun. They came together in a vast efflorescence of plant and fish, marsupial, lizard and fowl at Windorah and gathered at Coopers Creek for the glide to Lake Eyre. There the full Diamantina drained its own rich veins to complete the filling of the ancient sea. The Barcoo grunter, Cooper Creek catfish and golden perch came to vigorous life and multiplied. The new plant life drew great flocks of grey teal, ibis and pelicans, and here and there a great cloud of budgerigars blocked the sun. Everywhere nature broke old banks and built the new.

I am writing in the days after an electoral flood has dug new channels for Australian political history. This is much more than the periodic alternation between Labor and conservative that has occurred since Alfred Deakin's fusion of Free Trade and Protection in the first decade of Australia's national story. The liberal social democratic political culture that underwrote the orderly management of conflict and broadly based prosperity for most of my life had dried to a string of shrinking waterholes. Now the electoral flood has broken the debris and spread nutrients out of the old channels. The scene is set for new life.

What will we make of it?

My book *Reset* described Australia's choice: restoration of Australia, or post-pandemic Dog Days. The pandemic greatly disrupted economic life, but we cannot simply return to what preceded it. We can't return, because a trillion dollars of debt, the breakdown of the global trading system and our trading relationship with China, and the world's determination to phase out use of coal and gas – all of these block old paths.

In any case, we shouldn't want to return to the Dog Days. They were years of persistent unemployment and rising underemployment, of stagnating productivity and real incomes of ordinary Australians. And years of Australia working against its own national interest in humanity's struggle with climate change.

The reset offers great opportunity if we embrace it. Sustained full employment with rising incomes requires strong growth in investment and output in our export industries. Old paths are blocked by changes in our international environment, but these can be bypassed if Australia uses its natural and human resources to become the Superpower of the emerging low-carbon world economy.

The election came at a time when the accumulation of old mistakes and failures was interacting with international challenges to lay traps at every turn. I have been plotting the accumulation since early this century: in 2005, I spoke about the Great Australian Complacency of the early twenty-first century. This was a phenomenon of the developed democracies and was manifested in the global financial crisis in 2008 and 2009 – in which Australia managed to avoid recession with well-judged fiscal and monetary expansion here and in China.[1] But the complacency weighed more heavily over time in Australia. In my 2013 book *Dog Days*, I foreshadowed a period of stagnation, caused to a significant extent by the dominance of vested over national interests in public policy. Among much else, the increased weight of private interests led to the destruction of carbon pricing that had worked effectively to reduce emissions.

The news was not all bad. In the early winter of 2015, I explained in a public lecture at Adelaide University that despite the failures of energy and climate policy, Australia's exceptional renewable resources opened an opportunity for large economic benefits as the energy Superpower of the zero-carbon world economy. I expanded this story in *Superpower* in 2019, describing how Australia could do well economically despite the climate policy distortions in the years from 2013 if we participated in the global movement to zero emissions.

In 2021, immediately after the pandemic recession, a return to broadly based prosperity required new approaches to fiscal, monetary and labour policy for full employment. It required new approaches to taxation and social security for equity and productivity growth. It required the embrace of Australia's Superpower opportunity to reach these ends with a reasonable amount of external debt.

The biggest barriers to Australia choosing a better future are not in our stars, but in ourselves. We have accepted the degrading of our democratic political culture. We have let vested interests dominate public policy choice. We have abandoned knowledge, broadly shared through our community, as the foundation of successful public policy. We have kidded ourselves about how well we are performing as a democratic polity, society and economy.

The Coalition government in its last two years made important progress on full employment – in talking about it as an objective, and in gearing fiscal and monetary policy to that end. It was an important achievement that unemployment fell to the lowest rate for half a century. Otherwise the Morrison government chose Dog Days, with its failures on climate and energy and on general reform to lift productivity and living standards.

The May 2022 election changed what is possible. There is a large majority in the House of Representatives and a majority in the Senate for action on climate change and committed to the integrity of our democratic institutions and the restoration of knowledge to a central place in public policymaking.

The new parliament and immediate energy crisis help Australians to stop kidding ourselves. Our problems are large, and restoration will be difficult. The active and passive builders of the Dog Days, and the beneficiaries, will not fade away without a fight. The problems left from the Dog Days and the pandemic will provide them with allies and weapons. Let's not delude ourselves that a prosperous democratic future can be secured without good thinking, hard work and the preparedness of citizens motivated by the public interest to take strong stands on principle.

The tightest traps set by Coalition climate and energy policies were sprung by the Russian invasion of Ukraine and the dislocation of Russian trade and the extraordinary increase in global gas and coal prices that followed. The shocking increases in prices immediately after the election in May generated some calls for the government to proceed more slowly with support for increased renewable energy supply, and some for it to accelerate movement. Analysis presented in this book demonstrates large net benefits in acceleration, accompanied by judicious concern for power reliability. This will be the defining issue of the Albanese government's first term.

The new prime minister, Anthony Albanese, said in the election campaign and in his acceptance speech on the night of 21 May that his government would build Australia as a renewable energy Superpower. In the early stages of the Russian war energy crisis, the prime minister held his ground at the first meeting of federal, state and territory energy ministers on 8 June. The government would reduce emissions by 43 per cent on 2005 levels by 2030, and increase the share of renewable energy in the national electricity market to 82 per cent by 2030. On 16 June, Australia advised the UN Framework Convention on Climate Change that it had strengthened its commitment to emissions reduction by 2030 to 43 per cent.

This book is mainly about how the new government and its successors can capture Australia's opportunities. It examines in detail the traps on the path to building the Superpower. At the time of my Climate Change Review in 2008, contributing our fair share to the necessary global effort was expected to involve significant early costs. Moderate costs over the first half of this century would be recouped with a large surplus from damage avoided after that. Since 2015, it has been clear that the early economic costs of strong Australian and global action were much smaller, and the early economic benefits much larger, than anticipated in 2008.

In *Superpower*, I wrote that, 'We need to build a bridge on which Australians can walk over that chasm, from policy incoherence to hope and opportunity.' Australians occupying different places on the political spectrum and locations in Australia have been walking across the bridge. The Superpower narrative was prominent in the New South Wales Coalition government's nationally important legislation on energy in November 2020 and in the South Australian Coalition government's statement on climate change policy in the following month. The Queensland Labor premier in October 2021 said that she was taking steps towards making her state a low-carbon energy Superpower. In providing funding for a renewable hydrogen project in January 2022, the WA government proclaimed its goal of becoming a green hydrogen Superpower. Tasmania was recognised by the World Wildlife Fund as Australia's low-carbon Superpower leader in November 2021. The government of Victoria has taken steps to capture the opportunity.

The federal Coalition failed to face the challenge. While it spoke of opportunity, it saw the transition as a matter for the distant future. Its policy was built around a false dichotomy of 'technology not taxes'. Commonwealth finance was allocated abundantly to the 'gas-led recovery' and hardly at all to supporting the emergence of Australia's future strengths. Australia consistently supported Russia, Saudi Arabia and other countries opposing strong climate action in international discussions, against our strategic partners in North America, Europe and the Southwest Pacific.

There was an important shift in Australian community and business discussion of climate change over the years of Coalition government.

Even in 2013, exit polls revealed more Australians favouring retention of carbon pricing than introduction of Tony Abbott's 'Direct Action'. Over time the shift was large enough to make inaction on climate change electorally dangerous. In the year of the Glasgow climate conference, many Australian businesses declared their intention soon to operate with zero net emissions. A few have announced plans for major zero-emissions projects to export electricity, hydrogen and zero-emissions goods. Leading farm business organisations, including the National Farming Federation and Meat and Livestock Australia, have goals of zero net emissions by or before mid-century. Virtually all substantial businesses and business organisations accepted the goal of zero net emissions by 2050 before its adoption by the Coalition government – although many large producers and users of coal and gas argued that it was impractical to do much before the 2040s or to actually achieve zero by 2050.

The Coalition was never serving Australia's national interest with its resistance to strong global action. The Coalition government's formal acceptance in late 2021 of net zero by 2050 without credible commitments to strong intermediate steps was too little, and dangerously late. It showed an end point on a road map, but there was no motor to move Australia onto and across the bridge and no steering mechanism to keep it on course. The weak 2030 target, avoidance of commitments on methane reduction or the phasing out of coal use in power generation and absence of policy to achieve 2050 goals left us lonely among the developed democracies at the Glasgow meeting. Even slow forward movement was bogged in the accumulated political debris of recent history. That was one important reason why the electoral flood of May 2022 swept the Coalition government away.

Australia's success in building the Superpower is important to global success. It is important to sustaining full employment with rising living standards in Australia. And it is critically important to

Australian security. It was centrally important in the prime minister's and foreign minister's meetings – in May, June and July 2022 – with the heads of government of Japan, India, the United States and Indonesia, at the heads of government meetings of NATO in Europe, the G20 in Indonesia and the South Pacific Forum, and in the foreign minister's multiple visits to countries in the South Pacific. Global failure in holding temperature increases to around 1.5°C would destabilise Australia's Indo-Pacific neighbourhood. Continued failure on climate change would undermine Australia's leadership of cooperative approaches to development and international policy in the South Pacific. And full participation in the global mitigation effort would greatly strengthen Australia's relationships with security partners in North America, Northeast Asia and Europe.

The new government's policies allow participation in the global effort. It would be very much in Australia's interest to go further and to play a leading role. This overview essay sees building Australia's zero-emissions future in three stages. The first is dealing with the immediate energy crisis caused by the Russian invasion of Ukraine. The challenge here is to ameliorate damage to Australians' living standards without diverting the polity from building the Superpower. The second stage is to make a large start on decarbonising the economy. The implementation of the government's electoral policies would provide the main action over the first three years. This could be accelerated and expanded in other areas where there is clear opportunity to reduce costs and improve reliability of energy supply without contradicting electoral commitments. The third stage is the expansion of the new zero-emissions Australian economy into the Superpower role in global trade. The new industries would be an important element of the economy by the late 2020s and grow strongly from there.

The three stages can proceed simultaneously and strengthen each other.

Nine years of kidding ourselves

There are few major issues on which the general shape of the Australian national interest is clearer than climate change. Nor is there a major issue on which discussion of the national interest has been so badly damaged by political confrontation, unconstrained projection of private interests, and real and contrived ignorance. We have spent a decade kidding ourselves about the nature of the challenge and the effectiveness of our response.

The atmosphere for discussion of the issue was very different one and a half decades ago. My first official Climate Change Review was commissioned by the six state and two territory governments in April 2007 and then joined by Kevin Rudd's Commonwealth government in December. There was wide support for policy that would be in the interests of most Australians – in the national interest. The review was transparent in its analysis. Working papers were posted on the web and a draft report was discussed in town hall meetings that booked out quickly and filled to overflowing in all mainland capitals and many provincial centres. The review was supported by the federal Opposition led by Malcolm Turnbull, and discussed with him and relevant shadow ministers on a number of occasions. The Greens adopted a sympathetic, constructive and critical approach.

There was a genuine search for the national interest on an issue that was understood to be difficult and important.

Coal companies and their lobbyists and other emissions-intensive businesses which thought that their commercial interests would be advanced by delaying or deflecting action sought to influence policy outcomes in less transparent ways. Such action is an inevitable part of policymaking in a democracy. A democracy's quality is defined and its national success determined by how it handles such sectional pressures. The policy proposals that emerged were imperfect but good. Legislation embodying the main recommendations passed the House of Representatives in late 2009. Opposition support was necessary because the Labor government could not secure passage in the Senate with the Greens alone. The government focused its political diplomacy on the Opposition, not the Greens. After some compromises on compensation for emissions-intensive industries, the Opposition gave its support for passage in the Senate.

Then, on 1 December 2009, just before Prime Minister Rudd was to depart for the United Nations conference in Copenhagen, the leader of the Opposition was defeated by a single vote in the Liberals' party room. The government's climate change legislation was the main policy issue. Tony Abbott replaced Malcolm Turnbull. The Opposition withdrew its support for passage of the legislation through the Senate. Famously, the Greens opposed the legislation. That did not seem to be relevant, as they did not hold the balance of power in the Senate. When the bells rang, two Liberal senators crossed the floor in sympathy with former leader Malcolm Turnbull to vote with the government. Maybe the Greens' votes would have mattered after all and the government should have been engaging them in real discussions.

That was the end of broad political cooperation in search of the national interest on climate change. We will gradually learn whether the May 2022 election marked another turning point.

My second Climate Change Review was commissioned after the 2010 election by the Commonwealth government as advice to the Multi-Party Climate Change Committee. This committee and my second review were established within the agreements to form the minority Labor government led by Julia Gillard. We worked on a compressed timetable. The recommendations released in mid-2011 were improved by the maturation of thought by myself and others in the years from 2008.

The work of the Multi-Party Committee and my second Climate Change Review through the second half of 2010 and first half of 2011 were fiercely contested by the Abbott Opposition and its praetorian guard in the News Corp media. The public discussion was poisoned by bitter division, extreme use of social media by commercial interests (including through the new phenomenon of 'astroturfing'), abuse, threats of violence, intimidatory demonstrations against the prime minister and distortion of facts by the dominant media company. Political division provided cover for vested interests seeking to deflect or block effective action. Major business organisations, including the Business Council of Australia and the Minerals Council of Australia, invested heavily in opposition to the proposed legislation.

While the external political process was divisive and ugly, the work of the Multi-Party Committee was one of the most productive and constructive episodes of policymaking that I have observed in nearly half a century of close association with policy in Canberra. Government, Green and Independent members' constructive participation and Prime Minister Gillard's deft leadership led to crisp decisions on establishing carbon pricing, the Australian Renewable Energy Agency, the Clean Energy Finance Corporation, the Climate Change Authority and the Climate Council. Amendments to taxation and social security laws to protect the real incomes of low- and middle-income Australians were designed and passed into law, returning to households most of the large revenue from sale of carbon emissions permits (about $7 billion per annum initially and expected to rise over time). The role of the Renewable Energy Target (RET) in encouraging 41 terrawatt hours of grid-scale renewable energy and acceleration of installation of rooftop solar was confirmed. The Carbon Farming Initiative underwrote a new rural industry by allowing landowners to sell legitimate carbon credits. Carefully calculated numbers of permits were to be issued free to trade-exposed emissions-intensive industries, ensuring there was no advantage to international competitors that did not face similar constraints, under rules that maintained incentives to reduce emissions.

Regular review by the Productivity Commission would ensure that the assistance for trade-exposed industries remained appropriate. One weak spot in a strong policy outcome was multi-billion-dollar payments – smaller than in the 2009 policy package but without justification in the public interest – to coal generators as compensation for carbon payments that were actually passed on to consumers. The 'compensation' of coal generators with future payments for carbon permits was paid as a lump sum at the beginning to cover all future carbon pricing. None of the payment was returned to the government when carbon pricing was repealed in 2014.

The Clean Energy Future package passed the House of Representatives and the Senate with minimal amendment. The main features were in operation from 1 July 2012. A fixed price for three (later two) years was calibrated to the price of permits in the European Union's emissions trading scheme, in preparation for integration into the European system on 1 July 2015 (later brought forward to 2014). As it happened, the European price fell in the months leading up to the introduction of the Australian scheme, so that the initial fixed price ($23 per tonne of carbon dioxide equivalent) was criticised for being above the European. The European price in mid-2022 is now about five times the initial Australian fixed price.

The Clean Energy Future package operated very much as predicted. Carbon emissions subject to carbon pricing fell as anticipated – at a rate that would have given us zero net emissions by 2050 if that trajectory had been maintained. Large policy innovation rarely goes smoothly, and economic effects usually diverge from those anticipated by the models. This large innovation was different: reality followed the modelled outcomes. The leader of the Opposition, Tony Abbott, predicted the wipeout of Whyalla and other provincial manufacturing centres. The managers of the Whyalla steelworks took advantage of the incentives built into the assistance for trade-exposed industries to make a profit by shifting from higher-emissions hematite to lower-emissions magnetite ores. The leader of the National Party, Barnaby Joyce, predicted a several-fold increase in the price of lamb. The actual increase was negligible – as expected, since agriculture was not covered by carbon pricing. The percentage increase in the Consumer Price Index was as anticipated. The purchasing power of low- and middle-income Australians was protected by the taxation and social security adjustments. The scene was set for expanded trade with Europe in substitutes for emissions-intensive goods and services.

Carbon pricing was abolished on 30 June 2014, the day before trade with Europe was scheduled to begin.

It became the intention of the Abbott government after the 2013 election to repeal all elements of the Clean Energy Future package, as well as the Renewable Energy Target. Repeal was a close-run thing and failed on everything except carbon pricing. In the end, repeal needed the support of four Palmer United Party senators. Palmer had only three until a surprise result in a late rerun of the WA Senate vote provided a fourth. The Palmer Senate list had been cobbled together at the last minute and its campaign funded by $20 million drawn from a nickel processing company with large carbon-pricing liabilities (and teetering on and later falling over the edge of insolvency). Clive Palmer initially said that he would support the whole of the Abbott repeal package. A meeting in Canberra with former US vice-president Al Gore changed his mind – on everything except carbon pricing, where Queensland Nickel's interests were large. The Palmer United Party group secured passage of the Abbott repeal of carbon pricing and started to disintegrate soon after the crucial Senate votes. Not long after that, it didn't exist at all.

Thus is history made.

It is over a decade since we have been able to discuss in an ordered and informed way Australia's national interest in climate change policy.

The incoherence through these years has been called the climate wars. Their legacy of distorted perceptions of reality will take time to unravel.

It is worth reminding ourselves of some of the falsehoods that were given credence. Some are so deeply ingrained they are still accepted as truth today, even by people who recognise the value of climate action.

One is that carbon pricing was the subject of great political antipathy and the cause of the Labor government's defeat in the 2013 election. Rather, the government fell because of its bitter internal divisions and two changes in prime minister in three years.

A second is that electricity prices were much higher during the period of carbon pricing than ever afterwards. They were expected to be, with protection being provided for trade-exposed industries and households. But that is not actually how things turned out. Higher prices for electricity were the other side of the coin to the revenue collected by the Commonwealth through sale of permits and returned to households as increases in the tax-free threshold and social security payments. Wholesale prices fell briefly after the repeal of carbon pricing in 2014, but then rose to well above the levels of the carbon-pricing period. By the time of the 2022 election, they were several hundred per cent higher in the most coal-dependent states, New South Wales and Queensland. Unlike the price increases from carbon pricing, there is at present (July 2022) no carbon revenue to compensate households for the much larger price increases in 2022.

Through the climate wars and during the 2022 election campaign, the Coalition government claimed that South Australia and Victoria – the states that had gone furthest and fastest with expansion of wind and solar energy – were experiencing higher electricity prices after the closure of coal generators. The truth is that the closure of the Northern power station in South Australia and Hazelwood in Victoria pushed up prices for a short while. Then, with the expansion of renewable energy supply, these states came to have lower prices than Queensland and

New South Wales. They continued to do so through the crisis in electricity markets that began in May 2022. This was a new situation for South Australia, which had always had higher power prices than the eastern states. Dylan McConnell in Chapter 4 provides details.

The Coalition government argued that Australia was making good progress in reducing emissions from 2005 levels under its policies. In truth, when the temporary reductions in emissions from drought-induced falls in sheep and cattle numbers and from COVID-19 reduction in mobility and economic activity are stripped out, there was no reduction at all from 2005 in total emissions outside electricity generation. And the electricity reductions were the result of policies that the Commonwealth government had sought to abolish, or of state and territory initiatives that were criticised by the Commonwealth.

The Morrison government presented 'modelling and analysis' to the 2021 UN Climate Change Conference in Glasgow to demonstrate that Australia had policies to achieve zero net emissions by 2050. This was an elaborate exercise in self-deception. Close inspection of the document showed that Australia would be nowhere near zero net emissions by 2050 with established policies and technologies. What progress there was resulted from the assumption that businesses and households would act as if there was a carbon price – when actual policy was to exclude carbon pricing.

It was a fraud. It was seen by our strategic partners and competitors alike as a fraud.

In addition to persuading foreign governments that Australia was on a path to net zero emissions, the 'modelling and analysis' document was meant to convince Australians who had been opposed to climate action that zero net emissions by 2050 was a desirable objective. It made two correct points in pursuit of the latter. It said that the commercialisation and increase in scale of deployment of zero-emissions technologies would substantially increase Australian incomes. True.

And it said that the cost of capital in the whole economy would rise, and investment and incomes growth fall, if we did not do our fair share in the international effort. True. But sadly the policies proposed would not have militated against these outcomes.

The Morrison government sought to persuade Australians that the policies it took to the Glasgow conference were in line with those of other developed democracies. In truth, the 2030 target was the weakest among developed countries, the proposed path to 2050 was incredible, and Australia failed to sign up to the sectoral emissions-reduction commitments of other developed democracies. US president Joe Biden's first challenge after Glasgow was to get his Build Back Better fiscal package through Congress. It contained additional support for climate action equivalent to about $57 billion in Australia. President Biden said in advance of Congressional consideration that its passage would show that democracy still works, at a time of competition with authoritarian political systems. Biden struggled to secure Congressional support for legislation to achieve his goals. Australia was seen as being unhelpful. The Biden White House privately was furious at Prime Minister Morrison's failure to move closer to the new US target of a 50 per cent cut by 2030. It remained furious as this book was in preparation in 2022. A senior US official told leading journalist at *The Age* and *The Sydney Morning Herald* Peter Hartcher, 'We were looking to a close ally for support at Glasgow on a key policy and Morrison gave us just f—ing bullshit'.

The Coalition government sought to demonstrate a conflict between reliability and low cost of power supply on the one hand, and expansion of renewable energy and reduction of emissions on the other. In fact, the government's reluctance to plan for a zero-emissions electricity system caused power to be more expensive and less reliable. It deterred investment in technologies to balance intermittent renewables, and also in cost-reducing solar and wind generation. The changes in policy, the frequent arbitrary interventions affecting investment in the power system (abolition of carbon pricing, attempted abolition then reduction of the Renewable Energy Target) and large-scale public investment in generation (the Kurri Kurri gas turbine) and storage (Snowy 2.0) without specification of the rules that would govern the operation of public assets inhibited private investment. This managed to reduce power reliability, increase prices and slow reductions of emissions, all at the same time. It contributed to the 2022 energy crisis.

The Coalition government's declarations about policy seriously distorted policy choice. 'Technological neutrality' and 'technology not taxes' became the mantras of resistance to action on climate change.

'Technological neutrality' meant there could not be any systematic favouring of lower – over higher – emissions ways of producing goods and services.

'Technology not taxes' introduced a false dichotomy between market incentives for reducing emissions and the technologies through which they would be reduced. It is old economic knowledge that technological improvement proceeds more rapidly if favoured technologies are subsidised or less desirable alternatives taxed.

With these mantras, the Coalition and its media and business supporters departed comprehensively from the conclusions of mainstream economics and also of market fundamentalists, including Milton Friedman.[2] Official discretion replaced market exchange in resource allocation. This was damaging not only to reducing emissions but also to economic efficiency and to the integrity of our political system. Australia returned to the rent-seeking political economy that had preceded the reform era from 1983 to early in the new century. It was a significant factor in the fall in productivity growth to the lowest rate since Federation.

Now we have a chance to think and act again in the national interest.

Defining the national interest

It is now widely, if not universally, accepted that there is a strong national interest in the world holding temperature increase to no higher than 1.5°C, and in Australia playing a proportionate part in achieving that outcome.

There are three main reasons for this. First, the damage from global warming to Australia is severe. Ours is the developed country that will be most damaged by climate change.

Second, supplying a decarbonising world with zero-carbon goods and services is good for Australian prosperity. It is the most promising path to the expansion of new Australian export industries, which is essential for full employment with rising living standards and a manageable amount of external debt. Expansion of established exports is blocked by the post-pandemic inhibition to movement of people, affecting markets for services; by the rest of the world's adjustment out of using gas and coal; and by the breakdown in the open global trading system and our trading relations with China. The Superpower opportunity provides alternatives to the old export industries.

Third, positive approaches to climate change mitigation are important to Australian security and international relations.

Australia has an overwhelming national interest in achievement of the agreed international goal of holding temperatures to 1.5°C, in being recognised as playing a constructive role in that effort, and in being accepted as a legitimate supplier of zero-carbon goods and services to other countries committed to achievement of zero net emissions.

A national interest trade-off with energy reliability and security?
We want reliable and low-cost energy. Do we compromise this if we play our full role in the global transition to 1.5°C?

The technologies are available now to provide low-cost reliable alternatives to fossil energy. We can look forward to low and reasonably steady prices for electricity after the completion of the transition to zero net emissions, but not before. Reasonably smooth transition requires innovation in policies and institutions. There are speed limits on transition, mostly associated with minimum planning timelines and the availability of skills and management capacity. Sound planning and policy can expand capacity and raise speed limits. Transition to net zero in electricity would take a dozen or so years from where we are in mid-2022 – if we make a strong and purposeful start now. The costs and risks of transition have to be accepted at some time. They are not reduced by delaying the start or prolonging the process.

A national interest trade-off with prosperity from gas and coal?

The world's movement away from use of fossil carbon and hydrocarbon is no small thing in the world, and especially in Australia. The modern global economy has been built on intensive use of fossil carbon. Solar energy and atmospheric carbon dioxide had been converted by photosynthesis and natural storage processes into coal, oil and gas over hundreds of millions, in some cases billions, of years. This gradually converted a carbonic into the oxygen-rich atmosphere that can support life like ours. Over the past two centuries, the fossil carbon has been depleted, to drive the machines of the newly industrial world and meet the expanding consumption demands of households enriched by economic growth. This has returned to the atmosphere more and more of the carbon dioxide concentration that had been stored as fossil carbon. The availability of this concentrated energy was important to the burst of incomes growth that revealed the tendency for human fertility to fall when living standards became higher and more secure. The decline in fertility, in turn, allowed labour to become scarce and real wages to rise. This was crucial in denying the Marxist prognosis that capitalist economic development would inevitably bring immiseration to people who had only their labour to sell. Without the boost from fossil energy, the countries that are now developed may never have climbed out of the Malthusian and Marxian trap that had blocked sustained increases in standards of living for ordinary people from the beginnings of human civilisation.

So take a bow, coal, oil and gas, and all of the people who have worked in producing and distributing them from the early decades of the industrial age. Humanity's ascent from poverty and ignorance would not have happened without you.

But the fossil carbon age has to end soon.

It has to end one day because large, accessible, concentrated fossil resources are finite. The ever-increasing demand for fossil carbon would drive ever-increasing costs of oil, coal and gas. This would constrain global economic growth. And it has to end sooner rather than later because we cannot afford the destabilisation of the equable climate in which human civilisation, including the modern economy, has been built and upon which it depends.

The finite nature of economically accessible fossil carbon resources does not mean that they would be exhausted at any early date. But the coal, oil and gas resources that have highest quality, that are most easily and cheaply extracted and that are most accessible to centres of demand are those used first and exhausted. Over time, the fossil carbon frontier shifts to more costly sources and world prices rise. We saw this in fast-forward in China from early this century. The world's most populous economy grew rapidly for two and a half decades, within a development model that used energy and other industrial raw materials intensively. World prices of oil, coal and gas rose by 400 and 500 per cent in the first decade or so of this century, giving Australia its China resources boom. The world responded by preparing to draw on higher-cost sources of fossil carbon – including coal from the Galilee Basin in deep inland Queensland and coal-seam gas in Queensland and New South Wales.

Without concern for climate change, there would have been an extended period of rising prices, increasing investment in alternative technologies and gradual diminution of the relative importance of coal, oil and gas. But from early this century, climate change policies in Europe and some other parts of the developed world led to greatly expanded research, development and commercialisation of energy, industrial and transport technologies that did not depend on fossil carbon. This taught us something new: in many areas of economic activity, these new technologies would supply important goods and services reliably at lower cost than the old carbon-based technologies. This allowed continued economic growth to be reconciled with stopping climate change, but also lifted what had appeared to be a constraint on global development.

The Stone Age did not end because we ran out of stones, or the bronze age because we ran out of copper and tin. They ended because alternative technologies were deployed that provided necessary inputs into human material wellbeing at lower cost than the old. But there is nothing in the laws of economics that told us in advance that renewable electricity or electric automobiles would be cheaper than using the old carbon fuels. We have discovered over the past decade that they are.

A significant proportion of today's global energy demand is now being met from renewable sources. We reached peak global coal use in 2014. Peak production of cars that burn petroleum came a few years later, in 2017. In the absence of the climate change imperative, the transition away from fossil carbon would have happened gradually, probably without major political response or intervention. True, Ned Ludd and his associates broke mechanical stocking frames in opposition to new industrial technology in late-eighteenth-century Britain. But no bill was placed before the House of Commons to end the young industrial revolution. Candlemakers and blacksmiths accepted their declining relevance and value quietly, as electric lighting and automobiles rapidly expanded their roles in the early twentieth century.

In contrast, the accelerated decline in the fossil carbon industries is resisted strongly in Australia, the United States and some other countries. Is this because the decline is driven not by impersonal market forces, but by people acting for the good of humanity, guided by many of the best and most knowledgeable scientific minds? *Homo sapiens* is a morally complex species.

Total electricity generation costs – both capital and operating costs – in most parts of the world are lower using combinations of renewable energy than fossil energy. In the best locations for wind and solar, first of all Australia, the operating costs alone of fossil energy in average market conditions exceed the total costs of renewable energy. This was true before the Russian war lifted coal, gas and oil prices. In Australia, renewable energy is much cheaper in mid-2022 after accounting for the storage, demand management and diversification that are necessary to make renewable energy available whenever it is needed.

While fossil carbon is important to development everywhere, it looms larger in Australia than in other developed economies. Australia has by far the largest and richest coal and gas resources relative to population and economic size in the developed world. Australia's annual production of coal per person (about 18.7 tonnes in 2020) at the beginning of the 2020s is about seven times larger than that of our nearest rivals: China (2.7 tonnes), Russia (2.7) and Indonesia (2.1). Our consumption of coal per person also stands far above any others: 5.1 tonnes in 2020, compared with Germany (3.1), Korea (3.1), China (3.0) and the US (2.2).

Australia's gas and coal resources are internationally tradable at a relatively low cost, and available in other countries at similar average prices to those in Australia. (Gas in Western Australia is an exception, where domestic reservation policies keep prices low.) Thus they provide little or no advantage to Australian energy-intensive industry. So the Australian cost of adjustment to the decline of coal and gas production arises from decline in production for export. The size of export markets depends on others' climate change policies and not our own. It follows that the decline of the Australian gas and coal industries would not be fundamentally altered if Australia stood aside from the global mitigation effort – except to the extent that Australia standing aside affects that effort.

The phasing down of coal and gas as sources of wealth for shareholders and senior executives and employment and incomes for other employees has been anticipated by much of the community for a long time – in detail for a decade and a half. It is human nature for people facing unwelcome change to hope that something turns up to alter the reality – and if they see avenues for influencing outcomes, to use them. That hope has been encouraged by Australian governments from 2013 to 2021 denying that we will achieve zero net emissions. It has also been encouraged by the appearance of longevity created by expansion of Australian production over this period. There has been a new burst of hope in the scarcity and high prices caused by disruption of Russian exports after the invasion of Ukraine. Hope has also been encouraged by disinformation from vested interests in the coal and gas industries, and the continued critique of global and Australian action by Australia's dominant private media group. Some political parties have encouraged and magnified ignorance and hope.

Australia will lose its old exports of coal and gas at a pace determined by global progress towards zero net emissions. To the extent that Australian action affects that pace, our strategic and economic as well as environmental interests unequivocally support placing our weight behind achievement of the agreed goals of the international community.

There is value for the Australian coal and gas industries as a whole in continuing coal and gas exports from established mines at a rate geared to declining global demand, but not in investing in new, or expanding old, mines. Other developed-country governments and companies have reduced investment in new capacity, creating high prices and profitability for Australian exports. That is good for Australian coal producers and for Australia, but it creates a risk that Australian producers are seen as free-riding on others if they invest in expanding productive capacity. The national interest would be served by Australian companies joining the restraint on new investment. Private investors in gas and coal companies operating in Australia would do well, generating large cash flows from high prices without capital expenditure.

In current circumstances, restraint on investment in new capacity can be left to private investors, policed by their shareholders and suppliers of debt finance. Mistakes could leave shareholders with stranded assets and higher costs of capital – and possibly with lower export prices. There is no case for Australian governments sharing that risk by providing fiscal support for expansion.

Summing up the national interest

We have a powerful national interest in the world holding the temperature increase to 1.5°C. We can improve the chances by putting our hand on the strengthening side of the global scales.

We have a powerful national interest in moving swiftly to zero emissions in the many activities in which the newer technologies have lower costs than the old. These include electricity generation and road transport. Fast transition can only be achieved with sound planning and stable policies that allow markets to work efficiently by compensating private investors for benefits their actions bestow on others. With these in place, achieving zero net emissions in electricity generation by 2035 is not only feasible, but will also lead earlier to stable low prices with high reliability.

We have a large interest in Australia being accepted as a good-faith contributor to the world effort on climate change, and in our goods and services being accepted as zero-carbon products in other countries.

That would be assisted greatly by Australia moving purposefully to zero net emissions from electricity generation by 2035. There is no advantage in delaying or prolonging the electricity transition. Australia has a large interest in securing value from the coal and gas industries as they inevitably phase down and out. That is best achieved by enjoying the cash from established mines as they are depleted, rather than sinking capital into new mines.

What's our fair share? Australia's place in the international effort

Meinshausen et al. in Chapter 2 show that developed countries need to achieve zero net emissions by about 2045 if we are to do our fair share in the global effort to reach the Paris and Glasgow conference objectives.

The 'fair share' language came from the 2008 Climate Change Review. That was when earlier reduction in emissions was thought to carry a large cost, so that commitments to reduce emissions amounted to sharing a burden. That is not the case now for Australia and some other countries.

The 'fair share' conception remains relevant in a different way. International trade in zero-emissions goods and carbon credits is going to play an important role in the global economy. Trade raises incomes in exporting countries and reduces costs of mitigation in importing countries. But trade in credits has to be built around agreed baselines. The exchange will only be accepted by trading partners if the baselines represent a reasonable assignment of emissions-reduction responsibilities among countries. Similarly, countries importing zero-emissions or low-emissions goods will only accept them from countries that have similar or lower emissions constraints.

To complicate the picture, developing countries are generally experiencing faster economic growth than developed. They are at a stage of development when emissions intensity is especially high. And they have fewer economic and technological resources to make the transition.

Hence the international community agreed that developed countries would lead the emissions-reduction effort, with developing countries following.

For all of these reasons, my Climate Change Review recommended adoption of 'modified contraction and convergence', whereby developed countries were required to converge from the present towards a common end point, and developing countries were required to commence the convergence when their per-capita emissions had reached the (falling) levels of developed countries. That still seems appropriate, within a general expectation that developing countries will achieve net zero emissions by the mid-2060s. Meinshausen notes that this approach allows countries with high emissions per person at the beginning – Australia having the highest of the developed countries – to absorb a disproportionate share of the remaining global greenhouse emissions 'budget'. That doesn't look fair to many developing countries. But participation by developing countries becomes more attractive if they can sell credits for emissions reductions in excess of the target. This will be a real prospect in countries with high-quality renewable energy and landscape carbon resources. Developing countries are in a better position to keep emissions low and to earn credits from staying below their baselines because they are able to apply directly the new low-emissions technologies without needing to dismantle established high-emissions energy and industrial systems.

The Glasgow meeting fell well short of commitments to limit temperature increase to 1.5°C. The sum of targets did not achieve net zero emissions by 2050. Nor did their 2030 targets put those countries which had committed to net zero by 2050 on course for that objective.

Glasgow was nevertheless the occasion for substantial progress. It saw the belated acceptance of net zero by 2050 by the one country that had broken the consensus of the developed democracies: Australia. Russia (2060) and India (2070) also made commitments to zero net emissions for the first time. All developed countries except Australia substantially strengthened their short-term commitments – mostly getting close to, meeting or beating the global benchmark of minus 50 per cent by 2030. The Glasgow Compact asked Australia and other countries that had not upgraded their 2030 commitments for Glasgow to do so for COP 27 in Egypt in 2022. It was agreed at Glasgow that unabated coal use would be phased down – with India, China, Russia and Australia deflecting pressure for stronger wording. A commitment to phase out power generation by the 2030s was joined by all developed countries except Australia and the US – with the US nevertheless agreeing to zero net emissions in power generation by 2035 in a bilateral agreement with China. President Joe Biden led in a new agreement to reduce methane emissions by 30 per cent by 2030. The methane pledge was joined by all developed countries except Australia.

Glasgow also set a framework for trade in carbon credits. The US–EU agreement on trade in aluminium and steel seeks opportunities for trade in zero-emissions goods, and to restrict import of high-carbon goods. That is likely to establish standards that go well beyond the metals and the countries that were its initial focus. The US–China agreement reached immediately before the Glasgow conference in 2021 commits both countries to cooperation in emissions reduction – again with large trade implications, and with China going beyond previous statements on methane, coal and structural change.

On some issues important to Russia, China, India and Brazil, Australia was the sole fellow traveller from the developed democracies. *The New York Times* described the line-up of teams in its summing up of the Glasgow Conference on 13 November 2021: 'some countries, like the United States and European Union, did step up their climate pledges under the Paris Agreement. But others, like Australia, China, Brazil and Russia, hardly improved on their short-term plans.'

A new UN report in April 2022 set the scene for the Egypt meeting. It noted that the world as a whole is lagging badly on reduction of emissions to hold temperature below 2°C, let alone 1.5°C. A late and slow start requires a faster finish. Global temperatures will continue to rise until net emissions are zero. The IPCC's 2022 report contemplates overshooting to 1.7°C and clawing back with subsequent negative emissions – from large and sustained reductions in short-lived greenhouse gases like methane, or secure sequestration of atmospheric carbon in the landscape. This is a risky strategy but the only viable approach to the Glasgow goal of 1.5°C if we wait too long to make decisive cuts in emissions.

A chance at redemption

We in Australia have spent nine years since 2013 kidding ourselves that it does not matter on which side of the global climate scales we place our weight. We matter at least as much now on climate as we have done on any international issue on which we sought to influence outcomes. Probably more. It has mattered that domestic political opponents of the US president have been able to point to one close ally of the US which is on their side. The meeting of minds between Prime Minister Albanese and President Biden on 24 May in Tokyo, immediately after the election, began a process of correction. We can make sure that we make up for the mistakes of Glasgow in Egypt in November 2022.

Five policy shifts would see us doing our fair share as a developed country at COP27 in Egypt. These are all consistent with the electoral commitments of the Australian government elected in May 2022. We must:

1. Establish credible policies to achieve zero net emissions by 2050. Chapter 2 of this book indicates that net zero by 2050 for the world requires developed countries to get there earlier. But for Egypt, showing that we are going to make it by 2050 keeps us in acceptable company.

2. Commit to early emissions cuts within the range of other developed countries. The Labor government's electoral commitment of minus 43 per cent by 2030 just sneaks in – similar to Canada and Korea, but below other developed countries. At the Egypt conference, confirmation that Australia was formally committed to minus 43 and had policies in place to get there, and expected that its policies would achieve minus 75 per cent by 2035 without formally committing to that outcome, would be seen as a strong contribution to the global effort. It would also be achievable in the course of building the Superpower.

3. Join the pledge of over 100 countries led by the United States in Glasgow to reduce global methane emissions by 30 per cent by 2030. Australian methane emissions come from two main sources: fugitive emissions from gas and coalmining and preparation for export and use; and ruminant animals, mainly cattle and sheep, in the farming and grazing industries. Australia would exceed its obligations under this pledge if it phased in over the period to 2030 – actually, even to 2035 – a requirement to remove, sequester or fully offset all fugitive emissions in the coal and gas industries. The requirement would cost the industry a small proportion of the additional profits from increased prices that have resulted from diminution of investment in coal and gas production elsewhere in the world as a result of climate change. It would cost an even smaller proportion of the unanticipated additional profits from the higher energy prices following the Russian invasion of Ukraine.

4. Join President Biden's support for phasing out power generation emissions by 2035.

5. Pay our fair share of the $100 billion per annum that developed countries have promised developing countries

for emissions reduction and climate change adaptation. Australia has been talking about stepping up its contribution to investment in renewable energy in Southwest Pacific developing countries in recent years. There are large opportunities for expanding this effort and extending it into Indonesia and neighbouring Southeast Asia, in ways that support the Superpower ambition. Geo-strategic competition with authoritarian states in the Southwest Pacific expands the reasons for decisive action.

The agenda for COP27 in Egypt has been set up for countries that did not commit to their fair share by 2030 at Glasgow to redeem their failures. Australia has a special place in the spotlight. It is in our national interest to match the performances and meet the expectations of other developed democracies.

Electricity prices and the Russian war energy crisis

The timing of the election of the Albanese Labor government has haunting parallels with that of the Scullin Labor government in 1929. Scullin won an overwhelming majority in elections on 12 October. The Cabinet was sworn in on 22 October. October 24 saw the Black Thursday sell-off on the New York Stock Exchange that heralded the Great Depression.

The energy crisis that hit the new government in its first days has the potential severely to disrupt the economic welfare of most Australians. This comes on top of the stagnation in real wages and living standards through the Dog Days of 2013–19 and the pressures of the pandemic recession. Without major policy adjustments, we will see the largest declines ever in real wages between 2022 and 2025, with gas and electricity prices contributing much of the decline. Poor reaction to the crisis could undermine the prospects of the new government.

The energy crisis also has the potential to knock Australian decarbonisation and the building of the Superpower off-course. However, by July 2022, the new government was saying that the crisis argues for accelerated decarbonisation rather than retreat.

The Australian energy crisis has its immediate origin in the lift in global gas and coal prices that followed the Russian invasion of Ukraine in late February 2022. The international increases found their way into Australian prices over the following few months. If the high international prices persist and institutional and fiscal arrangements are left exactly as they were when the government was elected, several per cent of Australian household income would be transferred to producers of gas and coal over the three years from 2022 to 2025. The inflationary effects of the fossil energy price increases are driving the Reserve Bank of Australia to increase interest rates. Falling real household incomes and higher interest rates introduce the risk of Australia moving into its second recession in three years, after an unprecedented three decades without recession.

The 2022 Australian general election was on 21 May. The Minister for Energy and Climate Change, Chris Bowen, was sworn in on 1 June with most of the Cabinet. At the ceremony in Parliament House, the new minister was receiving text messages from his office and new department about the gathering energy crisis. Wholesale gas prices were reaching unprecedented levels through eastern Australia. At times they were thousands of per cent higher on average than in the preceding year – from a base several hundred per cent higher than before exports of liquefied natural gas (LNG) from Gladstone in Queensland lifted low domestic prices to East Asian levels from around 2015. Coal prices had been rising since talk of the Russian war began circulating. Gas and coal pushed electricity prices into new territory.

On 30 May, after the swearing in of the new prime minister but before that of most Cabinet members, the Queensland cumulative gas price threshold over seven days exceeded the extremely high level that triggers the imposition of a price cap of $40 per Gj under market rules. On 7 June, price caps on gas were extended throughout eastern Australia. Wholesale electricity prices right through the National Electricity Market (NEM) were then several times higher on average than in preceding years. On 12 June, the Queensland cumulated electricity price threshold over seven days exceeded the extremely high level that triggers the introduction of a price cap of $300 per MWh. Two days later, price caps were imposed in other states.

At the capped gas and high coal prices, many generators could not operate profitably at the regulated maximum wholesale electricity price. The rules made provision for generators to be compensated for losses in these circumstances. The rules had not been applied before and there was uncertainty how the compensation would work. Generators began withdrawing from the market. Anxieties arose about shortfalls leading to blackouts. The Australian Energy Market Operator (AEMO) began directing generators back into the market. Direction attracted more certain compensation and was favoured and sought by many generators. More generators withdrew from the market. AEMO lost track of the generating capacity that was actually available. On 15 June it declared the market 'impossible to operate' and for the first time since the establishment of the NEM in 1998 suspended the electricity spot market in all regions. Normal operation of the market was tentatively restored on 23 June without the wholesale electricity price cap and confirmed on 24 June. Average prices rose again above what had been the cap.

There has been much talk of the perfect storm driving the crisis of early June 2022. The unusual weather gave way to the usual cold. The rate of breakdowns of coal generators returned to that expected of an ageing fleet. Mines affected by floods returned to earlier levels of output. Yet electricity and gas prices remained extraordinarily high. The high international gas and coal prices from the Russian war were enough for the crisis.

The crisis has varied impacts across the Australian continent. There is no problem of high gas and electricity prices for the 11 per cent of Australians resident in Western Australia. The Carpenter government in 2006 required gas exporters to place 15 per cent of production on the domestic market. Gas markets are well supplied at well below international prices. Low-cost renewable energy has been supplying an increasing share of electricity. WA coal is not exportable, so its cost is not affected by international coal prices. WA electricity-price increases continue at or below the consumer price index.

The problems are most acute in the two states in which coal remains overwhelmingly dominant in electricity generation: Queensland and New South Wales.

How electricity prices are set

There are five electricity regions, corresponding to the states, with the ACT part of New South Wales. Interconnection allows movement of power in response to price differences. There is insufficient interconnection capacity to equalise prices across the regions.

Retailers and users of power in each state offer to purchase quantities of power at specified prices for each five-minute interval. Generators in each state offer to sell quantities of power for the same intervals. The price for each interval is set so that the sum of offers to buy at or above that price equals the sum of offers to sell at or below that price.

I will now tell a stylised story of how prices are set through the interaction of supply of power from different technologies with total demand. I sacrifice some complexity for clarity.

There are three types of power generation. One type is renewable: solar, wind and hydroelectric. Most of the cost of all three is borne at the beginning – the capital cost of setting up the facility. For solar and wind, once the plant is in place, costs are no higher if the generator is delivering power to the network than if it is not. Indeed, there are costs in spilling power, so total costs are a bit lower if power is delivered. Solar and wind generators bid into the market at a price near or below zero. They bid the volumes that come from the sun and wind conditions of the time. They receive the price which clears the market each five minutes and not the low price that they bid.

Hydroelectric power is different. Water can be run through the generator at any time, so selling in one period at a low price reduces the amount of power that can be sold later when the price may be higher. If the price is low in the market, energy will be held back for later sale when prices are higher. In this way, hydroelectric power has the characteristics of stored power.

The second type of power, from coal, has substantial operating as well as capital costs. The main operating cost is the coal itself. If the coal is unsuitable for export, the cost of production determines the price of coal. If the coal is exportable, the value from exports determines the price at which it is made available to domestic generators. It is costly to change the output of a coal generator. Once coal is being delivered into the market, production is only likely to cease if it is expected that the price will be unprofitable for some time.

The third type, gas, has lower capital and higher operating costs than coal generation. It is more flexible, and so can respond more quickly in raising or lowering output of power in response to changes in market prices.

Demand for power varies through the day and the seasons. Demand through the grid varies with generation of small-scale solar in households and businesses: if people are generating more power themselves, they draw less from the grid.

Price is determined by the size of demand in each five-minute period, relative to supply from the various sources. If there is enough solar and wind to meet demand, when added to inflexible coal generation, wholesale prices are near or below zero. This now happens frequently in the regions with largest wind and solar energy capacity – about a third of the time in late 2021 in South Australia, and nearly that in Victoria.

If there is not enough solar and wind generation to meet demand for power in a region in some five-minute period, the price will be set by the next lowest-cost source of power. This is usually coal.

If renewables and coal generation together are insufficient to meet demand, gas generation fills the gap. Prices are then higher.

The average wholesale price depends on the relative proportions of time at which the price is set by renewables, coal and gas. Increasing the renewables share expands the proportion of the time in which renewables meets the whole of demand and prices are very low. More renewables also increase the amount of time when renewables or renewables plus coal together meet demand – keeping prices lower than when some gas power is required.

A contraction of coal supply increases the proportion of time when some gas generation is required. That increases average prices.

The story so far is about how prices are set at a particular time. The dynamics of price formation over time are more complex.

The closure of a major coal generator changes the balance of the several sources of power. We have had examples with the closure of the Northern power station in South Australia in 2016 and Hazelwood in Victoria in 2017. The withdrawal of large amounts of relatively low-cost coal power increased the proportion of time when high-cost gas generation was required. This led to a large immediate increase in average wholesale prices. After that, the supply of renewable energy increased by large amounts. This increased the proportion of the time when renewables set the price at very low levels. It reduced the proportion of the time during which gas set the price at high levels. The growth of renewables after the coal closures shifted average SA prices well below those of Queensland and New South Wales when they had once been consistently higher.

The result is a sawtooth pattern of price changes over time. Expanding supply of renewable energy (including from rooftops) reduces average power prices. This eventually leads to the closure of a coal generator. Prices jump, and then resume their downward slide. The profile is of a sawtooth blade with a downward slope.

The dynamic is disrupted if international coal and gas prices increase. These raise the level of the saw while leaving in place the profile of the saw teeth.

Eastern Australian electricity prices are vulnerable to increases in international gas and coal prices for as long as internationally tradeable coal and gas figure prominently in supply. They are vulnerable to periodic closure of coal-fired generators until the last coal generator is closed. When renewables supply almost all requirements, average power prices will be relatively low and insulated from international energy markets. The path to low and relatively predictable prices is an uneven one, with prices lifting in the aftermath of each coal closure, and with any increase in global prices of gas and coal.

In Chapter 4, Dylan McConnell provides empirical detail on forces driving recent wholesale electricity price changes.

Fixing the energy crisis

The energy crisis of mid-2022 is driven by dislocation of global markets from the Russian invasion of Ukraine. The dislocation of fossil energy markets from the Russian war is likely to persist after the ending of military action. Distrust and trade restrictions will remain. A global recession – a possibility – would alleviate the problem but only for as long as the recession continues. Regrettably, a weak global economy may ease the energy crisis for longer than is comfortable. The full pass-through of international into Australian power and gas prices will take more than two years, as households and businesses come off old and enter new contracts, and regulatory agencies take account of these lags in their pricing decisions.

The price problems come from both coal and gas. There are two ways to protect household living standards. One is to stop higher international prices flowing back to domestic prices, by driving a wedge between them. The wedge could come from forcing enough gas and coal onto the domestic market to meet demand at a price well below international levels, or by placing a levy on exports. The other is to provide compensation payments to consumers from Australian public revenue, drawing on the increased profits from higher prices accruing to the gas and coal exporters.

A wedge between export and domestic prices: Domestic reservation

Strong domestic reservation policies could be introduced into eastern Australia, along the lines used successfully in Western Australia. Existing legislation enacted by the Turnbull government would need to be strengthened if it were to return domestic gas prices to those before the Russian war. There would be a Senate majority for new legislation.

Another possibility is for the three LNG exporters voluntarily to expand supplies to the domestic market by enough to push prices back to pre-war levels.

A wedge between export and domestic prices: An export levy

A 'Russian war price normalisation levy' could be applied to exports of coal and gas at a rate that would remove the increase in prices above the average level before the Russian war (say, the average of the year before the war). This would hold export prices received by companies after paying the levy to the (relatively high) pre–Russian war levels. Domestic prices would fall by the amount of the levy. The levy would be applied by the ATO shipment by shipment. Old contracts with pricing set before the Russian war would only attract a levy to the extent that the contract itself linked price to international oil or gas prices – as most do. It would not affect the integrity of sales contracts. The structure of the levy would see it phasing out automatically as prices fall back over time to pre-war levels.

The revenue from a Russian war price normalisation levy would rise gradually over time, for as long as the abnormally high international gas and coal prices remained. What to do with the revenue?

Compensatory payments to households and businesses would not be required, as prices for gas and electricity would return to levels before the Russian war. The Queensland government, with royalty payments increasing with price (to 40 per cent on high prices for coal and 12.5 per cent for gas), may stake a claim for compensation. That case would be qualified politically by Queensland residents being full participants in the lower power and gas prices. The revenue could be returned to coal and gas producers in a way that is unrelated to current export revenues – for example, in proportions equal to gas and coal companies' shares of export revenue in the year before the Russian war. This would hold down prices paid by domestic users of coal and gas while leaving producers with most of the Russian war windfall. Or it could be used to pay off some of the trillion-dollar Commonwealth debt.

The different histories of coal and gas may warrant different responses: measures (regulatory or voluntary) to increase domestic supply of gas; and a Russian war price normalisation levy for coal. The combination would end the continuing transfers of incomes from households to coal and gas producers in eastern Australia. It would end the large gas and electricity contributions to high inflation, inflationary expectations, demand for wage increases and pressure for higher interest rates to curb inflation. The fall in gas and power prices could be a decisive circuitbreaker in the Reserve Bank's concerns about the inflationary cycle driving increases in interest rates through the middle months of 2022.

A tax on windfall profits to fund compensatory payments to gas and electricity users?

In the early months of the energy price crisis, former Treasury secretary Ken Henry called for a windfall profits tax on the increased earnings from the increase in coal and gas prices. Visiting Nobel Laureate Joseph Stiglitz supported the idea, calling it a 'no brainer'. The Conservative UK government introduced a similar measure in the northern spring of 2022.

A windfall profits tax could protect the real incomes of Australians by taking a moderate fraction of the windfall gains from the increases in gas and coal prices – much less than the Australian Fraser Coalition government's crude oil excise in similar circumstances in the mid-1970s.[3] However, unlike measures to drive a wedge between international and domestic prices, it would not reduce the inflation indices monitored by the independent Reserve Bank for its decisions on interest rates. For that reason, I would favour domestic reservation or a Russian war equalisation levy over a windfall profits tax.

The Superpower destination

Superpower: Australia's Low-Carbon Opportunity described the large role that Australia can play in supplying zero-emissions products to the rest of the world. This can substantially reduce the cost and increase the prospect of the world achieving its climate change goals. And it can provide Australia with the increases in investment and production in export industries that will underwrite full employment with rising real incomes. Australian zero-carbon exports can be large enough to reduce substantially the costs of decarbonisation in Germany and other densely populated industrial countries in Europe, and in China, Japan and Korea in Northeast Asia. Large enough to have some effect in Southeast and South Asian countries with limited renewable energy resources. Australia would also play a crucial role in easing global supply bottlenecks for a range of energy transition minerals, and in providing carbon offsets where other countries are not able to remove emissions from some activities at moderate cost.

The Morrison government made hydrogen export and green aluminium and green steel central to its Technology Roadmap. But for the former government, the zero-emissions developments were for the distant future. Emissions-intensive industries within Australia felt comfortable in continuing with current technologies into the 2040s before making major changes.

The May 2022 election result liberated a wider spectrum of the Australian community to talk honestly about climate change and the transition to zero net emissions. More elements of the business community came to face reality. Parts of the community that had been facing in the right direction strode forward with greater confidence. For some, the new government's commitment to use the Emissions Reduction Fund's safeguard mechanism to reduce emissions in major industrial activities forced the realisation that some early change was necessary. The final version of AEMO's Integrated System Plan, released on 30 June 2022, fleshed out a 'Hydrogen Superpower' alongside its 'Step Change' phase-out of almost all coal generation by the mid-2030s. Tony Wood and colleagues at the Grattan Institute on 3 July 2022 released 'The Next Industrial Revolution,' a report on utilisation of the Superpower potential.

The Superpower opportunity is built on five endowments that Australia has in much greater abundance (relative to population and economic size) than any other country. These are renewable energy resources; land suitable for growing biomass for zero-emissions industry and sequestering carbon; the world's largest supply to international markets of minerals requiring large amounts of energy in processing; energy transition minerals, which will be scarce and valuable over the next several decades; and the legacy of knowledge, skills, institutions and infrastructure from the old resource industries.

Let us take a closer look at each of them.

First, we have distinctly the richest per-capita endowment of wind and solar in the developed world, and as good as in any developing country. Most electricity will be generated directly from wind and sun in future, and most energy will be provided to industry as electricity. Hydrogen from renewable electricity will become the main process of reducing iron ore to metal, and a major source of materials for chemical industry. Nuclear and other zero-emissions alternatives available in globally significant quantities will be more expensive than renewables based on Australian wind and solar.

A country that has the lowest-cost solar and wind and broadly competitive storage has the world's lowest-cost energy in the zero-carbon world. Australia has opportunities for balancing the intermittency of solar and wind power at lower cost than in the general run of developed countries. Alongside hydroelectric power and pumped hydro storage in Tasmania and the Snowy Mountains, there is potential for this in many locations in the Great Dividing Range and nearby in New Guinea. There is abundant potential for pumped hydro storage in the voids of depleted mines. Australia's leading position in renewable electricity has made it a leader as well in use of both grid and decentralised battery storage. Its established and new Superpower industries have much potential for using power flexibly in response to availability and cost. Long-distance transmission can contribute reliability by joining different solar and wind conditions across a continent. There are high-quality carbon geo-sequestration sites adjacent to rich natural gas resources for flexible thermal generation.

Almost all the costs in renewable energy generation and balancing are in capital expenditure. This is very different from industries drawing energy from fossil carbon, where a high proportion of costs are from purchasing oil, gas or coal year by year. Capital has lower costs in developed than in developing countries. This gives Australia another advantage in a zero-carbon world, over developing countries that have renewable energy resources of similar quality.

Australia's second source of comparative advantage is exceptionally large per-capita endowments of woodlands and other land suitable for sustainably growing and harvesting biomass at low cost. This gives us comparative advantage in producing biomass to replace oil, gas and coal in the chemical industries – for export as industrial raw materials, or for domestic processing. It also makes us potentially a large-scale exporter of carbon credits to countries which are unable to achieve zero net emissions from domestic resources at less than prohibitive cost.

Third, we are the world's largest exporter of mineral ores requiring energy for conversion into metals. We are also potentially a highly competitive producer of biomass-based materials requiring large amounts of energy for conversion into plastics, plant-based foods and other chemical manufactures. Iron, aluminium and silicon smelting are the largest examples of energy-intensive processes in which there are economic advantages to location in Australia in a zero-carbon world. There are many others.

Fourth, the world's movement to zero net emissions greatly increases demand for a range of minerals. These include aluminium and copper, which were important in the old economy, and are even more so in the new. There is greatly increased demand for what have become known as 'critical minerals'. Following Sandiford in Chapter 3, we prefer the term 'energy transition minerals'. Australia is exceptionally well endowed with them. They include silicon (as an oxide in sand or quartz) for solar panels and computers; a range of metals for wind turbines; a range of minerals for batteries and electric vehicles (copper, cobalt, nickel, manganese, magnesium, graphite, lithium, vanadium, titanium and other rare earths).

Fifth, the skills, business experience and institutions that supported Australia's export of minerals, energy, agricultural and forestry products in the old economy are valuable in the new. This is also true of much physical infrastructure, including high-voltage electricity transmission lines (once used to move electricity from the coalfields to manufacturing and urban locations and now available for transfers of renewable energy back to the old coal-based industrial centres) and ports. The legacy of the fossil carbon economy will be a large Australian advantage in the early years of establishing new industries.

Australia also had the richest per-capita endowments of coal and gas, so why didn't similar advantages for processing industries appear in the old fossil carbon world economy? There were times when they did, although on a smaller scale than in future. High energy prices and increased concern for air pollution in Japan in the late 1970s and 1980s saw aluminium processing established in Newcastle in New South Wales, Gladstone in Queensland and Portland in Victoria. Within a decade or so, Japan went from being the world's largest producer of aluminium after the Soviet Union to a country relying almost entirely on imports. Australia became the largest aluminium supplier to Japan and, for a while, largest exporter to the world. There was also substantial investment in chemical manufacturing in southern Australia in the last third of the twentieth century, based on low domestic gas prices. Other non-ferrous metallic minerals were largely processed in Australia in the twentieth century: lead, zinc, copper, nickel, manganese.

But processing is competitive. Producers can only expect a normal competitive return on investment. It is not like mining, where the state allocates exclusive rights to a private company, which can then generate rents from any superior-quality mineral resource. The rent value can in principle be collected by the state that owns the resource, but if it is not, it generates exceptionally high rates of return for private investors.[4] In Australia, processing was rarely as attractive to investors as simple mining. Processing became even less valuable relative to simple mining when minerals prices rose with the China resources boom in the first dozen years of the twenty-first century. Mike Sandiford provides details of this in Chapter 3.

Gas-based manufacturing in eastern Australia lost its cost advantages comprehensively with the LNG processing and exports from Gladstone from 2015.

Domestic gas prices rose to export parity levels – while US gas prices fell as the government there restricted exports as supply from shale increased.

Australia is the world's largest supplier of metallurgical coal as well as of iron ore into international markets. This might have been a source of advantage in production of iron metal and steel. It wasn't, as international transport costs for metallurgical coal represent a small proportion of value, and prices in importing countries are as low as or lower than in Australia. Judicious planning for two-way bulk transport of iron ore and coal would have improved the economics of processing in Australia. Mining companies never thought it warranted the effort. WA governments waived the domestic processing commitments that had been conditions of allocating iron ore mining rights.

In the years in which the aluminium smelting industry was established in mainland Australia, state-owned electricity businesses had privileged access to high quality thermal coal resources. They were able to use low-cost state capital to build generators, and transmission systems to link them to smelters in the industrial ports – in the case of Victoria, across Melbourne to Portland in the southwest corner of the state. The preferential allocation of low-cost thermal coal to electricity ceased with the privatisation (in Victoria and New South Wales) and corporatisation (Queensland) of generators. Ownership of coal in many cases was separated from power generation. The cost of coal in electricity generation in New South Wales and Queensland rose to export parity levels.

The new Australian advantages in minerals processing are much larger than the advantages in the fossil carbon economy ever were, and they are sustainable. Low prices for electricity and industrial inputs from Australia's renewable energy resources will not be removed by international trade.

Renewable energy can be economically traded across the seas, but international movement is expensive relative to the value of the products. Exports of renewable energy will only supply high-value uses that are not subject to international competition. Unlike Australian coal, they will not be imported to process Australian raw materials in competition with processing in Australia.

Electricity can be transmitted from Australia to Indonesia, Singapore and the adjacent Asian mainland. Australian renewable energy can be converted into hydrogen and liquefied for export, or into ammonia and other carriers for recovery of hydrogen at the points of import or use. International transport costs for zero-emissions electricity and hydrogen are so high that the products will cost more than twice as much in import destinations as in their place of origin in Australia.

It will be much more economically efficient to convert Australian mineral ores into iron, aluminium, silicon and other metals using Australian renewable electricity in Australia than in the rest of the world.

The advantages for countries with poor endowments of renewable energy and landscape carbon resources drawing zero-carbon goods from Australia are illustrated by Ligang Song in Chapter 6. China has committed to achieving zero net emissions by 2060. That will not be easy. China has large wind, solar and hydroelectric resources. They are thousands of kilometres to the west of the main centres of modern economic activity on the east coast. It has large land resources for tree plantations. It installs nuclear generators at lower cost than other countries. But China's resources for zero-emissions and negative-emissions production are small compared with its population and economic size, and easily exhausted. Imports of zero-emissions industrial inputs can fill the gap.

This iron and steel story has parallels in the Chinese fossil energy economy today. China has the world's largest resources and production of coal and substantial gas and oil, but it is also the world's largest importer of fossil carbon.

In Germany, the best areas for solar power produce half as much from the same equipment as the best in Australia. The best solar resources in Germany are inferior to the worst in Australia. Moreover, land is scarce and has high value in established uses. Solar and wind generation already occupies much land. The dislocation of Russian coal and gas supplies following the invasion of Ukraine has increased focus on hydrogen in Germany, as in the rest of western Europe. But limited domestic renewable energy resources are fully occupied meeting old decarbonisation goals. There has been increased German and European interest in Australia as a source of hydrogen. There is growing realisation that the early stages of iron processing are better conducted at the sources of iron ore and low-cost renewable energy.

Which industries for the Superpower and how big?

The largest single opportunity for reducing global emissions and raising Australian incomes is shifting Australian iron exports from ore to metal. The technology for direct reduction of iron ore using hydrogen is proven. Its use is most advanced in Germany and Scandinavia. It is commercially viable in the best Australian locations now at mid-2022 prices for metallurgical coal in the absence of a carbon price, or at the average metallurgical coal price over the past decade at the European carbon price. Europe is a natural early focus of sales if Australia is recognised as a legitimate participant in European low-carbon trade. Japan and Korea offer similar opportunities. China – producing and using more than half of the world's steel and producing much more than half of the world's primary iron – is the most important market in the longer term.

Magnetite ores are more easily concentrated into the high-grade iron oxide suitable for currently favoured hydrogen-based direct reduction processes. Research and development in China is extending use of hydrogen-based direct reduction to hematite ores. There are magnetite deposits in many locations in all Australian states and the Northern Territory. Many are small – but large enough for commercial ironmaking where electricity transmission and other infrastructure are available at low cost. The immense hematite resources in the Pilbara region of Western Australia currently dominate Australian production and exports, and will be converted into iron metal for export at a later date.

Ligang Song explains in Chapter 6 how Chinese demand for steel is likely to peak in the late 2020s and then gradually decline. The proportion of iron metal for steelmaking drawn from scrap steel will rise from low levels (about 15 per cent in 2021) towards the levels of developed countries over the next several decades (30 per cent in 2025, to about 60 per cent in 2050, compared with 70 per cent in the US today). So demand for iron ore in China will gradually fall from now, to less than half current levels in 2050. Mixing pure iron ingots from direct reduction of iron ore with hydrogen raises the quality of steel made from scrap in electric arc furnaces. This facilitates the transition from imports of iron ore to iron metal as electric arc furnaces using scrap play a larger role.

The decline in steelmaking in China will be accompanied by an increase in Southeast and South Asia. The young steel industries in these regions can rely from the beginning on imports of iron metal to support the later stages of steelmaking. Development in African and Latin American countries is more likely to be served by regional iron ore and renewable energy.

The conversion of scrap and pure iron into steel in electric arc furnaces is a highly electricity-intensive process (although less so than reduction of iron ore to metal using green hydrogen). This encourages a proportion of Australian iron metal exports being steel ingots.

Economics argues for shifting almost all Australian iron ore exports to iron metal over the next three decades. The economics is favourable for making a start on a small and medium scale now in the most favourable locations. Large-scale developments require new transmission and other infrastructure, and with good policy and management will emerge in the early 2030s and predominate in the 2040s. At current prices, conversion of two-thirds of Australian iron ore exports to iron metal with one-sixth being taken through to steel ingots would roughly double the value of Australia's immense iron ore exports.

The second-largest immediate opportunity is conversion of exports of bauxite to alumina (pure aluminium oxide) and aluminium (metal) ingots. Australia once converted most of its bauxite into alumina before export. The proportion shipped as unprocessed bauxite has increased in the twenty-first century. Australia is overwhelmingly the economically rational location for conversion of bauxite into alumina and alumina into aluminium metal in the zero-emissions world.

Both alumina refining and aluminium smelting are currently highly emissions-intensive in Australia, with the exception of aluminium smelting at Bell Bay in Tasmania. The most likely path to decarbonisation of alumina refining is through electrification and shifting to renewable electricity supply. Decarbonisation of aluminium smelting is largely secured by switching to renewable electricity. Australian mainland smelters are belatedly examining the switch. They would have lower costs if they switched as soon as their coal-based contracts expire. The combination of high-quality renewable energy, linked by transmission systems established for the Collie coal power stations, and adjacent alumina production, make southwest Western Australia the world's most favourable location for the next generation of aluminium smelters. The central and north Queensland industrial ports will also be strong candidates after the strengthening of transmission links to high-quality renewable energy resources in the near inland.

Mike Sandiford in Chapter 3 describes the opportunities for new mining associated with a seven-fold increase in global demand for energy transition minerals over the 2020s. In the zero-carbon economy, these are most economically exported in processed form. Australia is an established major supplier of lithium, copper, nickel, manganese, titanium and tantalum. It is a niche supplier of high-quality silicon. As Sandiford notes, Australia could become a major supplier of cobalt from mine wastes at Mount Isa and elsewhere. New mines and processing facilities are being established for vanadium, graphite and magnesium in a number of Australian locations.

High-grade silicon is crucial to the expansion of global solar power. China dominates silicon metal production – in 2021, about 70 per cent of the world's 8.5 million tonnes, valued at nearly $40 billion – and a higher proportion for grades used in PV panels. The crucial inputs for processing sand or quartz are electricity (more electricity-intensive than aluminium smelting) and pure carbon (anthracite, or char or charcoal from biomass). The carbon requirement will shift to biomass in the zero-carbon world. Higher grades attract a large premium.

China is not the economically natural home for the major part of global silicon production. It lacks comparative advantage in any of the necessary inputs: renewable energy, bio-carbon and high-grade silicon oxide. Economic realities in a zero-carbon world will cause it to stop exports and to import a proportion of its domestic requirement. Restrictions on export in late 2021 and early 2022 reflecting government commitment to constrain domestic carbon emissions were the main driver of a trebling of world silicon prices in subsequent months.

Australia is abundant in all the raw materials required to produce silicon metal. As such, it is well placed to play a major role in meeting the immense growing global demand for this for solar PV systems and computers. Today the one established producer, Simcoa at Kemerton in southwest Western Australia, accounts for about one half of one per cent of global production. It was located near Collie in 1987 to be adjacent to coal power at reasonably low prices by global standards, and it makes charcoal from Jarrah offcuts from sawmills. The supply of timber is diminishing, so a considerable volume of anthracite is imported from Columbia at high cost. In the future, bio-carbon resources may be drawn from pyrolysis of sustainably harvested suitable acacias and eucalypts.

Land for growing and sustainably harvesting biomass for energy and industry will be scarce and valuable in the zero-emissions world economy. Isabelle Grant in Chapter 7 notes the economic case for converting biomass to high-quality char (carbon) and biogas and bio-oil (hydrocarbon). These will be called upon to replace oil, coal and gas in a wide range of industrial processes. Bio-oil will be an important source of fuel in zero-emissions long-distance civil aviation. It is easily stored in facilities built for petroleum, so may be a source of long-duration thermal-power generation for an Energy Reserve that provides reliability as the proportion of intermittent solar and wind power increases (see Appendix 1.1).

Carbon fibre is a superior substitute for a range of manufacturing inputs. It can be many times as strong as a similar volume of steel with much less weight. Production requires a pure carbon source – now fossil carbon, but bio-carbon in the zero-emissions world – and immense amounts of energy. Australia has large advantages for production of the carbon-fibre raw material. Carbon Revolution, operating from the Deakin University industrial research hub Carbon Nexus in Geelong, has won an important place in global supply of carbon-fibre automobile wheels. The low weight makes the material important in high-performance vehicles, and also in electric vehicles where weight affects vehicle range. Australia's comparative advantage in making carbon-fibre products may extend a considerable way down the value chain from the energy-intensive and biomass-intensive stages.

In all countries, some emissions are extremely expensive to eliminate. These will require offsets from activities generating negative emissions.

Negative emissions will also be important if, as is likely to be the case, future generations decide that carbon dioxide concentrations in the atmosphere need to be reduced further once we have reached zero net emissions. Increasing carbon in plants and soils is the most promising technology for this. There are some opportunities for using biomass for energy and industry, and capturing and storing associated emissions in geological structures, rather than returning them to the atmosphere. This is known as bio-energy carbon capture and storage (BECCS).

While use of renewable energy in local manufacturing is likely to be commercially more attractive than direct export of electricity and hydrogen (in liquid form or as ammonia), the latter will still be important. Australia is likely to become a substantial exporter of electricity to Southeast Asia. The Sun Cable project is being developed to transport solar electricity generated in the Northern Territory under Indonesian waters to Singapore. It is looking in the first instance at supplying about 15 per cent of current Singapore requirements. The submarine cable will be laid across Indonesian maritime territory. Over time, Indonesia is likely to see advantage in drawing some power from this source. Once available in Singapore, Australian zero-emissions electricity can be transported into and through mainland Southeast Asia through established and expanding transmission systems.

Frank Jotzo in Chapter 5 discusses how green ammonia will be exported from Australia to replace high-emissions ammonia – mostly used in fertiliser and explosives. Total world consumption is around 180 million tonnes, of which a tiny proportion is in Australia. High-emissions production of ammonia from natural gas and coal has been concentrated in China, Russia and the Middle East. Australia currently uses about 2 million tonnes per annum of urea, the ammonia-based fertiliser. The one local producer, Incitec Pivot in Brisbane, meets about 10 per cent of Australia's requirements. It has announced that it will shut its Brisbane operations in late 2022. High global gas prices, especially after the Russian invasion of Ukraine, and Chinese restrictions on exports of energy-intensive and emissions-intensive products, have lifted urea prices to several times average historical levels in real terms in 2022 and generated anxiety about security of supply (see Susannah Powell, Chapter 8).

Ammonia and hydrogen will be produced for export to countries with poor renewable energy endowments. The main early markets will be densely populated developed countries in Northeast Asia and Europe. The Asian Renewable Energy Hub in the Pilbara is being designed to use about 26 GW of solar and wind capacity in producing 1.6 million tonnes of green hydrogen or 9 million tonnes of green ammonia for export. This would generate and use almost half as much electricity as Australia's NEM. Other ammonia and hydrogen export projects are being developed throughout Australia.

If Japan and Korea secure half their zero-emissions energy requirements domestically, and Australia secures the same share of Japanese and Korean zero-emissions energy imports as for coal and gas, it would export about 40 million tonnes of hydrogen equivalent per annum – as electricity, hydrogen, ammonia or embedded in zero-emissions goods. It may be realistic after the Russian war for Australia to aspire to a share of European markets half that of Japanese and Korean markets – so, 20 million tonnes per annum. If Australia secured one-tenth of its Japan–Korea share of the Chinese market for zero-emissions fossil carbon substitutes, exports of hydrogen would be as large as to Korea and Japan. South and Southeast Asia and bunker fuel for international shipping in the southwest Pacific, Asia and some to Europe might each also be about 10 million tonnes. The bunker ammonia would replace petroleum as an international shipping fuel, which currently accounts for almost twice the carbon dioxide emissions as Australia. Total exports would be in the order of 120 million tonnes per annum of hydrogen-equivalent.

Current Northeast Asian and European import plans do not speak of imports in these volumes. But the plans are in the early stages of development and have yet to absorb the implications of the Russian war for Europe, or anything like the full implications of zero net emissions. The initial REPowerEU plan for the European Union envisages imports of about 10 million tonnes of hydrogen by 2030, and strong growth after that. Separately, the port of Rotterdam alone is developing the infrastructure for imports of 4.6 million tonnes of hydrogen by 2030. The demonstration of the economic advantages of imported renewable energy in hydrogen, hydrogen carriers and embedded in goods will rapidly expand understanding of the advantages of much higher levels of foreign trade.

With the Superpower industries, Australia would be a high-wage economy. The only new trade-exposed manufacturing industries that did not depend on energy or biomass that would make economic sense would have other special advantages in the Australian environment. What might they be?

There would be huge local demand for high-voltage transmission cables. There would be advantages in making them close to aluminium and steel production and to where there is a strong market. There is currently an acute global shortage, especially for direct current. Most modern cables have extruded aluminium conductive cores, so would benefit from use of molten metal from an adjacent smelter. The aluminium core is surrounded by a steel case. An adjacent electric arc furnace converting green iron and scrap into special steel would have advantages. Transmission cables could be a constraint on development of the Superpower in the absence of local production. Production alongside an established smelter and a new electric arc steel furnace would have large opportunities in the international market as well. This could be a large industry in Gladstone, Newcastle or Portland. Newcastle has the initial advantage of substantial local demand for steel ingots.

Batteries? Maybe. That's to be tested by entrepreneurial effort. Local supply of all the local materials reduces supply-chain risks, but otherwise provides only limited comparative advantage. It is some advantage to have a large, dynamic local market, providing manufacturers with feedback on innovations in design and specifications.

Solar panels and wind turbines, for which there will be immense local demand? They are more labour-intensive, so probably not.

Exceptional specialised knowledge and skills are a major source of comparative advantage for some products. The Carbon Revolution carbon-fibre wheel plant in Geelong is an example.

How much power, how much investment?

The National Energy Market supplies about 180 TWh of power per annum, representing about 80 per cent of national electricity use. AEMO's 'most likely' scenario in its Integrated System Plan, called 'Step Change', sees the NEM growing to about 220 TWh in 2030 after reductions for energy efficiency and a large increase in rooftop and other decentralised solar. The increase supports electrification of household, industrial and transport activities that currently use oil and gas. 'Step Change' has NEM demand for power increasing to about 320 TWh by 2050 with acceleration of electrification, about half each from wind and solar. That would require capital expenditure on generation of about $225 billion at the prices of late 2021, or an average of about $8 billion per annum. The proportion of standalone residences with rooftop solar would increase from about 30 per cent in 2022 to half in the early 2030s and 65 per cent in 2050. Rooftop and other decentralised power would expand three times by 2030 and five times by 2050, making a substantial contribution. Electric vehicles would steadily and rapidly increase their share to about three-fifths by 2050. Decentralised batteries and EVs would supply a majority of the 46 GW of dispatchable power to balance the intermittency of wind and solar in 2050, with the rest coming from grid-scale pumped hydro and battery storage (16 GW), established hydroelectric (7 GW) and gas with capture and storage or offsetting of emissions (10 GW).

The 'Hydrogen Superpower' scenario in AEMO's 2022 plan projects the requirements of the electricity system if Australia emerges as a major exporter of hydrogen. It envisages NEM power use about one-third higher in 2030 and more than three times as large as the 'Step Change' in 2050. The total additional capital expenditure on solar and wind generation would be about $700 billion by 2050, or $32 billion per annum, taking the total to about $40 billion per annum. There would be additional storage and other balancing requirements, with decentralised storage playing a proportionately smaller role, and diversification of solar and wind resources a larger one.

This is impressive work by AEMO, lifting Australian horizons of what is possible. It demonstrates that far from threatening the reliability and affordability of power supply, faster expansion of decentralised generation and storage and its integration into electric vehicle use can improve both.

The full Superpower opportunity would require much more investment than AEMO's 'Hydrogen Superpower'.

Partially replacing current fossil carbon use in the rest of the world with imports of hydrogen and its carriers from Australia along the lines discussed above would require the equivalent of about 120 million tonnes per annum of hydrogen by 2050. That would draw on over 6000 TWh of power – over thirty times the electricity traded through the NEM today – or over six times the volume envisaged in the AEMO's Hydrogen Superpower scenario. The AEMO Superpower is best seen as the modest proportion of electricity supplied through augmentation of established regional electricity grids. Most of the Superpower activity would be in new nodes of power generation and use, as described in the discussion of the Sungrid in Appendix 1.2.

Where trading partners are using imported renewable energy or hydrogen to replace carbon-intensive processing, the economic case for processing in Australia would be compelling. The iron metal processing described above would require about 2200 TWh of power for conversion of iron ore to metal, and about 75 TWh for production of steel in electric arc furnaces. Together they would absorb nearly thirteen times as much power as is currently traded through the NEM. The replacement of exports of iron ore by metal would add to total demand for Australian electricity: all the power requirements would be drawn from Australia, whereas only a proportion – and outside Japan and Korea a small proportion – of hydrogen supplies would be from Australia.

Electricity requirements for zero-emissions refining and smelting of all Australian bauxite production would exceed current through-put in the NEM. Silicon processing has the potential to make larger demands on electricity than aluminium. I have not sought to quantify the requirements of producing and processing carbon fibre, energy transition minerals and biomass-based chemical manufacturing, but they are large.

Meeting the full Superpower opportunity would require in the order of 10,000 TWh of annual power generation.

Supply constraints on building the Superpower

Capturing a modest proportion of the Superpower opportunity would place immense demands on Australian political, administrative, community and economic resources. How much is feasible and over what time frame? And what can be done to increase our capacity?

Meeting the hydrogen, iron and other minerals processing potential described above would require about fifty times current NEM capacity by 2050, from over 3 TW of solar and wind capacity – or ten times the requirement of AEMO's Hydrogen Superpower scenario.

Current technologies and costs would suggest capital expenditure exceeding $6 trillion in today's costs, including associated high-voltage transmission and other mechanisms for firming intermittent renewable energy. Over two trillion dollars would be required for investment in manufacturing assets.

The simple arithmetic makes it clear that if Australia embraces the Superpower opportunity and is accepted as a full partner in the global effort, the constraint on the growth and scale of the Superpower is going to be domestic resources and not international demand. There will be pressure on labour skills, capital, management and corporate capacity, and government planning and regulatory capacity.

Labour will be scarce and valuable. With investment in Superpower export industries, we can choose to have full employment with steadily rising real wages. In this economy, educating and training our people well has great value. Creating jobs in itself does not. Getting the most value from a fully employed labour force is the essential task in building the Superpower.

The Superpower industries generally are highly capital-intensive, with the partial exception of biomass production and carbon farming, which will substantially increase demand for labour in rural Australia. Australia will need to focus strongly on technologies that have high labour productivity, alongside the education and training of Australians. Within this fully employed economy, the labour of every Australian worker is scarce and valuable. The valuable skills of Australian workers should not be wasted in unproductive protected industries. Neither should they be wasted in building new fossil-energy-based projects that are destined to be stranded in the zero-emissions world economy.

We will face important choices on immigration. I explained in *Reset* the advantages of a moderate level of immigration focused on the permanent migration of people with highly valuable skills. Higher levels of immigration, as in the China resources boom and the Dog Days that followed, would support higher rates of total economic growth and expand Australia's contribution to the global mitigation effort. But real wages would continue to stagnate or to grow slowly. That is likely to be incompatible with continued support for the economic strategy. It may eventually be accompanied by challenges to our democratic constitution – as such a strategy was in resource-rich Argentina a century ago, and as wage stagnation has been in the United States during the early twenty-first century. Better to seek broad and sustained community support around a moderate rate of immigration – the half per cent annual increase in population from immigration of the decade of rapid productivity growth in the 1990s, rather than the 1 per cent and occasionally more of the China resources boom and the Dog Days that followed.

Would the capital markets support investment of over eight trillion dollars in restructuring the Australian economy – over three times the current annual value of Australian production? The investment will be made gradually over several decades. An economy growing steadily at 2.7 per cent per annum – 0.7 per cent per annum from growth in the labour force (mainly from immigration), 2 per cent from capital and total factor productivity growth together – would be nearly two and a half times as large mid-century as it is now. Eight trillion dollars is around 5.5 per cent of output and income over the three and a half decades to timely completion of the Northeast Asian zero-carbon transition. Our domestic savings are much larger than that, but we will want to be investing in other things as well. Our immense superannuation wealth will need to be attuned as closely to investment in expanding export-oriented parts of the Australian economy as it is now to investment in natural and other monopolies at home and abroad.

The cost of capital will be critical. The global tendency over the past decade for private owners of income to want to save more than they invest will keep global real interest rates historically low. Australia will easily attract its share of that wealth for investment in the Superpower if we develop clear and stable policies and strategies, within a steadily growing economy with full employment, low inflation and a manageable level of public debt.

Private funding comfortably financed the comprehensive restructuring of the Australian economy to meet the opportunities of the China resources boom through the decade and a half from about 2002. ABS data show that investment in mining rose from 1.2 per cent of GDP at the beginning of the century to a peak of 7.9 per cent in June 2012, before easing back to 4.5 per cent in 2015. Other investment was fairly steady over this period. If the policy settings are steady and favourable to investment in the Superpower economy, it is not unreasonable to think of 5.5 per cent of GDP being invested in a sustained restructuring of the economy to meet new opportunity. With nominal GDP expected to be about $2.5 trillion in 2022–23, this would support investment of over $135 billion per annum in the Superpower at the beginning. With the economy growing by about 2.7 per cent per annum, it would support investment of over $300 billion per annum in the new economy at the time of completion of the Northeast Asian transition in the 2050s. Accumulated investment in the Superpower would be around 8 trillion in today's dollars by 2055–56. The roughest arithmetic says that we could support most but not all the investment required to realise the Superpower's potential. The endpoint would see China and Southeast and South Asia near complete decarbonisation. The completion of the Superpower transformation in the decade or so after the middle of the century could be crucial to those countries' success.

All these numbers will change in the course of the Superpower transformation. High costs of zero-emissions energy in importing countries may spur energy-saving innovation in consumption and production. We may see more renewables embedded in goods and much less as hydrogen. Structural change on this scale may be judged by Australians to be excessive. Australians may choose to save and invest more of their steadily increasing incomes. Governments may get their public finances in good order and invest more in the economy of the future, causing productivity, incomes, savings and investment in the Superpower economy to increase more rapidly than we have supposed. Continued reductions in the costs of the new technology may – and probably will – greatly reduce the amount of capital required. But we have shown that it is reasonable to expect the funding to be available to build the Superpower with today's technology and patterns of economic activity, over the years in which Australia's Northeast Asian neighbours will have to achieve zero net emissions.

Established regulatory and business institutions are poorly equipped to handle change on the required scale. New planning mechanisms will be required in federal, state and local governments. These will be the foundation of broadly based community understanding and support for sustained structural change on a massive scale. In the United States, President Joe Biden invoked the Defense Production Act for the first time since the 1940s, to secure large-scale mobilisation of goods that are important in the climate and energy transition. The Australian Super-power transformation is proportionately an order of magnitude larger than the US transition to net zero emissions. It will require intense national focus on identifying and removing impediments to success.

We will need more ambitious, dynamic Australian businesses, alert to the scale of opportunities and to the need for innovation in managing change on Superpower scale. We will need more ambitious, dynamic Australian businesses, knowledgeable about and comforta-ble with conditions in their own country. We will need ambitious and dynamic Australian and international businesses, able to turn away from today's rent-seeking Australian economy towards acceptance of risk as the price of reward.

3 TW of capacity is a lot of wind turbines and solar panels. If half is solar, it would cover 3.75 million hectares or 37,500 square kilometres,

or nearly half a per cent of the land area of Australia. If these were located at the seven nodes of the Sungrid and those in the three most remote locations – on the magnificent solar resource near the Tropic of Capricorn – were twice the size of the others, each southern solar region would require 3750 square kilometres – roughly a square with 60-kilometre sides. That is shockingly large. That will not all happen at once, and there will be time to assess and reassess as local communities live with the new opportunities. Renewable energy development on this scale would not be remotely possible in Japan or Korea or coastal or central China or Germany or elsewhere in Europe.

It is not out of the question in Australia, so long as it has the participation and support of local Indigenous communities. I dedicated my book *Superpower* to the Mutthi Muthi, Paakantiji and Ngyimpaa people, and to the big role that Indigenous Business Australia can play in building the new economy. The Iningai and its Turraburra community feature in Grant's discussion of landscape carbon in Chapter 7. In my experience, Indigenous Australians are deeply interested in making the most of this opportunity to prevent destruction from climate change of precious landscapes, and in making good use of landscape carbon, renewable energy and other new economic opportunities. They are important partners, and in many places leaders, in building the Superpower.

Frank Jotzo, in Chapter 5, draws attention to the critical role of environmental approvals and management in the development process.

There will be pressure on global supply chains to build the solar panels and wind turbines. In the early 2020s, these are overwhelmingly centred on China. China's manufacturing capacity will remain critically important to achievement of global net zero emissions. But China alone could not build anything like its current global share of renewable energy machine production, in the greatly increased production that is required in future. This would be the case even if its trading partners were comfortable about it doing so. We need an open international trading system for environmental goods, to support the emergence of new centres of production in the developing world, with large-scale investment from many developed countries and China. Australian trade minister Craig Emerson contributed to the development of an open approach to trade in environmental goods, including wind turbines and solar panels, in APEC a decade ago. The understandings reached then have held through the geostrategic, trade policy and pandemic disruptions of the past decade – with the important exception of US import restrictions directed at China during the Trump presidency. The new trade minister in the Albanese government, Don Farrell, has taken up this issue again at an important time.

There is sensitivity in the early 2020s about overwhelming reliance on Chinese inputs of pure silicon, rare earths and other energy transition metals. The concerns are partly geopolitical. There is a strong case on general insurance as well as geopolitical grounds for greater international diversification of supply sources. Australia holds out the most economically promising opportunities. Silicon and other inputs from Xinjiang have become controversial on human rights grounds in many developed countries. Australia can play an important role in providing a large-scale alternative source of these inputs – including for use in China in production of machines for global markets.

Policies for building a Superpower

Australian climate change policy starts from a long way behind, so we won't get to a good place quickly. It is our good fortune that our natural and human advantages are too large for past policy failures to disqualify us from participating now in the benefits of the new energy economy.

The new government's policies allow us to take some big steps, but to avoid false steps and unnecessary detours we should keep an eye on where policy needs to go in the longer term. It is worth going back to basics to get our bearings. Economic theory tells us to use one instrument for each policy objective. By trying to achieve many things by pulling one lever, or lots of things with each of many levers, we confuse ourselves and end up in a tangle. Here there are warning bells around the Energy Security Board's proposed 'capacity mechanism'.

It is a settled matter in economics that the community benefits from market exchange to allocate resources wherever there is effective competition. Good policy requires us to be clear on which activities can support competitive supply, and which have the characteristics of public goods for which supply through competitive markets is unsuitable. The advantages of competitive market exchange are so large that it is worth governments putting considerable effort into establishing conditions in which markets can operate effectively. But we cannot make a silk purse out of a sow's ear. In the energy sector, the network services are natural monopolies, and expectations that private decisions can lead to good outcomes for the whole society and economy are bound to be disappointed. Generation and retailing, on the other hand, can support competitive markets.

It is a settled matter in economics that if private participants in market exchange take decisions that confer large benefits or costs on others, beyond the private effects from exchange, government action is required if outcomes are to be favourable to the community. Community outcomes are improved through provision of a subsidy for a benefit, or tax for a cost. So we would do best with an economy-wide carbon price, linked to other countries' carbon markets by trade in emissions permits. That is for the time being ruled out because of the political history of the climate wars, so it is worth exploring other general mechanisms favouring zero emissions.

The value of explicit carbon pricing increases as we get closer to zero net emissions and it becomes more costly and difficult to remove the remaining emissions. The worst possible way of proceeding would be through interventions here and there to promote some zero-emissions outcomes, but which depend on officials' discretion. That builds the political economy of the rent-seeking society, with unsatisfactory economic outcomes and distortion of political systems. It was the way Australia proceeded with climate policy after the abolition of carbon pricing in 2014, contributing to our Dog Days.

In the electricity sector, the renewable energy target (RET) has been the main policy instrument driving increasing proportions of solar and wind so far. Its role would have been gradually rendered redundant if carbon pricing had been retained in 2014. But without the carbon price, the cessation of the RET in 2030 as currently planned would create a gap and a problem for Australia's clean energy future.

The other big external effect of market exchange is from innovation in new zero-emissions production. We would do well to have the public finances comprehensively reward such innovation. This recognises and compensates the risks and costs borne by those who move first to teach the rest of the community about various approaches to building the zero-emissions economy.

Here we focus on two primary objectives, for which we need to find effective levers. The first objective has two secondary goals.

The first primary objective is rapid expansion of renewable energy. This is essential to cutting emissions, bringing down power prices and building the Superpower industries. Meeting the government's aim of 82 per cent renewable electricity by 2030 would represent good progress but is easier said than done.

It involves two secondary goals. The first is to secure high standards of reliability in the electricity system as it becomes overwhelmingly supplied by renewable energy. The second is to supply the necessary network services. The network through which renewable energy is transmitted to users is a natural monopoly. Services will not be provided efficiently without government direction of investment. Being a natural monopoly, any private ownership must be subject to price regulation in the public interest – which is inherently difficult without risking systemic failures or introducing large incentives for wasteful over-investment.

Our other primary objective is rapid deployment of the new zero-carbon technologies in all the industries that comprise the Superpower. Here we look at the instruments through which each of the objectives can be met.

Primary objective: Rapid expansion of renewable energy

We can make a start without a carbon price, but would find it prohibitively difficult and costly over the later stages of the transition from the mid-2030s. A carbon price rising over time and linked to international markets will provide incentives to eliminate emissions that are particularly difficult and expensive to remove, or to develop genuine negative emissions to offset them. It will support mutually fruitful trade among countries that can produce carbon credits at lower cost, and with others that can remove all of their emissions only at much higher cost. It will establish the zero-emissions credentials of products traded on international markets. At that future time, the emissions trading scheme will not increase the cost of energy, because energy will be produced and traded without carbon emissions. It will not significantly increase the cost of transport, because there will be little combustion of fossil carbon in land motor vehicles. It will generally not greatly increase the cost of industry – and where it does, international competitors will carry similar costs to cover emissions that are particularly costly to remove. Carbon prices will be much higher than any previously contemplated in Australia, and provide powerful incentives for Australians to capture and store carbon in plants and soils in our landscape.

We will end up with a carbon price. But not soon. The Australian Secretary-General of the OECD, former Abbott and Turnbull government finance minister Mathias Cormann, is currently leading efforts to establish an emissions trading system across the developed countries. The OECD is likely to have some success. It will take time. US membership would require Congressional support, and therefore a turn of the US political wheel in a favourable direction. There is time for that – or for the wheel to turn against international cooperation on this and other things. Unlike preferential trading areas for the general run of goods and services, linking of emissions trading systems for a few but not all countries can be relied upon to bring benefits for trading partners as well as for the world as a whole. So there is value in an emissions trading system that joins many countries but not the US. Value, but not as much. California's cap and trade program is currently linked to other Western states in the US and Canadian provinces, and would be a valuable participant in an international carbon market in the absence of the United States as a whole.

Wounds from the climate wars will gradually heal. Progress in the zero-emissions economy and return to rational calculation of economic costs and benefits will one day make comprehensive carbon pricing possible. I suggest that we anchor our contemporary work with an expectation that Australia will be ready for an emissions trading scheme with deep international links by 2035. If it is not, we will stumble on with other instruments as best we can. In the meantime, we build on other foundations for production and trade in zero-emissions goods and services.

The Minister for Energy and Climate Change, Chris Bowen, said on 24 June 2022 that the best way to reduce electricity prices is to accelerate the growth of renewable energy supply. He is right. Achieving the government's stated objective of 82 per cent of electricity from renewables by 2030 would make a decisive difference in reducing prices.

Removing transmission bottlenecks between renewable energy zones and major load centres through the new Commonwealth government's Rewiring the Nation program is a necessary condition for success – but not sufficient. Dylan McConnell in Chapter 4 shows that investment in renewables has been declining in the 2020s. Beyond the transmission constraints, policy uncertainty has deterred investment. That can be removed by the new government settling on a policy framework.

High electricity prices are a powerful incentive for investment in renewable energy and in the storage and other firming of solar and wind power that must accompany it. But the high electricity prices from the Russian disruption of coal and gas trade are not expected to continue indefinitely, and until now price expectations have not justified investment sufficient to achieve 82 per cent by 2030. Therefore I suggest extending the end date of the Renewable Energy Target from 2030 to 2035, pending introduction of an economy-wide carbon price. Let's take a closer look at how the Renewable Energy Target works.

The RET was introduced at an unambitious level by the Howard government in the early years of the century and then greatly strengthened by the Rudd government in 2009. The Rudd version was designed to secure installation of 41 TWh of grid-scale renewable energy by 2020 – expected to be about 20 per cent of national electricity use. A renewable energy generator with capacity exceeding 100 KW receives large generation certificates (LGCs) in numbers corresponding to annual output. Smaller installations receive small generation certificates (SGCs), which attract lump sum payments reflecting the presumed value of reducing emissions through to the conclusion of the scheme.

As Dylan McConnell notes, the abolition of carbon pricing made the RET crucially important in maintaining momentum in renewables investment. Nevertheless, the Abbott government sought to abolish the RET and established the Warburton Committee to advise on its future. The chairman of the committee was on the public record as an opponent of action to reduce greenhouse gas emissions. The committee commissioned modelling that, to its surprise, but not to the surprise of economists familiar with the industry, indicated that the RET substantially reduced wholesale prices of electricity by increasing supply of renewable energy. In any case, after Clive Palmer's discussions with former US vice-president Al Gore, the government did not have the numbers in the Senate to abolish the scheme. The target was, however, reduced to 33 TWh in 2015 with the support of the Labor Opposition in the Senate.

The RET came to play an important role beyond the encouragement of investment in the targeted level of renewable energy supply. Surrender of an LGC to the Clean Energy Regulator provided evidence that a purchase of energy was backed by renewable generation. In the 2020s, this became the basis of a large voluntary market in renewable energy. Companies, governments and individuals wishing to demonstrate that they were purchasing green power relied on the RET system. This became increasingly significant, as companies and governments sought to demonstrate their commitment to zero-carbon supply chains. As a result, LGCs retain value even when the target has been notionally met by quantities of renewable energy generation.

The RET continues to serve important purposes. If it concluded in 2030 as currently scheduled, it would leave gaps and large problems. In the absence of a carbon price, there would be inadequate incentives for investment in renewables to achieve the government's 82 per cent. And there would be no mechanism for certifying that renewable electricity was supplied for a particular use for which green energy is essential. One could develop various mechanisms to serve these purposes, but none would be as straightforward as extension of the RET, and others are likely to be more expensive. Expansion of renewable energy supply encouraged by the RET is likely to continue to bring down average electricity prices for several years and at no budget cost. Extension of the RET is simple and uses a familiar instrument, the administration of which is well established. An extension of the closure date for the RET from 2030 to 2035, with appropriate adjustments of the targets, would close the gap in renewable investment incentives and certification of clean energy through the crucial years immediately ahead.

The extension would cover small as well as large-scale systems. Small-scale schemes attracting SGCs are used by households and small businesses. It was once said with some merit that support for roof-top solar was a subsidy for wealthy Australians, who could afford the capital outlays on a solar system. Now, with nearly a third of free-standing houses having small-scale solar, its removal would entrench the privilege of the wealthy and deny access to solar systems to people on average and lower incomes who had not been able to afford them when costs were higher. Institutional innovation is extending access to small solar systems to some people on lower incomes, including many in social housing.

Extension of the RET would demonstrate to the Australian electorate and to international partners that Australia had established mechanisms to ensure that it will supply 82 per cent of power from renewable sources and reduce total emissions by 43 per cent by 2030. This would make its commitments to zero net emissions by 2050 credible – distinguishing it from the incredible statements of the previous Coalition government.

Secondary goal: Reliability with predominantly solar and wind energy
It is a truism that solar energy is only generated when the sun is shining and wind energy when the wind is blowing. As the proportion of inter-mittent renewable energy in the system grows, greater effort is required to ensure enough power is available to meet demand at all times in all regions. That is reliability.

The prime minister and energy minister have described paying for reliability as buying insurance against the possibility of extreme damage from the breakdown of the electricity system. It is enormously costly for society if the whole electricity system ceases to deliver power for a period. The minister has drawn the analogy with water: rain doesn't fall all the time, but we have reliable access to water by storing it. Reliability can be provided by flexible use of conventional hydroelectric facilities; storage (batteries and pumped hydro being the lowest-cost at present); geographic diversification of solar and wind through long-distance transmission; thermal power designed to operate flexibly (in a net zero-emissions world, fuelled by gas with carbon capture and storage or carbon offsets, bio-gas or bio-oil, green hydrogen, or occasionally coal using mothballed generators with carbon offsets when back-up is required over a long period); and using power flexibly when it is scarce and expensive ('demand management').

How should we secure reliability? Price in a competitive energy market plays a role. When solar is abundant in the middle of a sunny day, the price is low and often negative in regions in which it meets a high proportion of demand. That provides an incentive to use more of it at that time, and to fill storage facilities. When the wind is not blowing after sundown when families are at home and active, the price is high. That provides an incentive to use less power at that time, to draw power from storage, and to use high-cost peaking generators. There are large opportunities for households and businesses to use power flexibly in response to variations in price. AEMO, in the most likely 'Step Change' scenario for its 2022 Integrated System Plan, envisages more than half of the required storage in a system dominated by solar and wind to come from decentralised batteries attached to solar generation and electric vehicles. The rate of deployment of decentralised batteries in the grid will be highly responsive to price changes and to fiscal incentives.

Markets will not reward private investors in storage and peaking assets fully for the high value they contribute by avoiding extremes of unreliability. Governments everywhere must take responsibility for preventing complete breakdown of energy systems. The system needs reserve generation capacity that can be called on to meet unexpectedly high demand or low supply, whatever the cause.

The balancing of demand and supply at all times in all seasons was always a challenge, even when coal generators supplied nearly all power and they plugged on through day and night. The challenge increases with higher proportions of wind and solar. The absence of planning for reliability as solar and wind generation increases leaves large problems now. This is the most dangerous, and probably the costliest, legacy of the climate wars.

The Energy Security Board's proposed 'capacity mechanism' was not relevant to the price crisis when it was raised for discussion in May 2022, but should be evaluated as a response to the need for reliability. The mechanism's design is to be finalised by late 2022 and operations to commence in 2025.

The ESB's proposal involved paying generators not only for power sold into the energy market, but also, separately, for having supply capacity available at times of strong demand – whether it is held in reserve for those times or engaged in normal generation for the market. Retailers and large users of power would have to buy 'capacity' alongside energy.

The original proposal was called 'Coalkeeper' by critics, as a major part of the payments would have gone to coal generators. It was expected to prolong the active life of coal generators. Since a major source of unreliability in recent years has been malfunction of ageing coal generators, it is not obvious that this would help.

The capacity mechanism needs to be compared rigorously with alternative means of achieving a high degree of reliability – on cost, compatibility with a competitive dynamic energy market, and time to implementation. The ESB's 'Consultation Paper on the High-Level Design of a Capacity System' said that higher prices are not the intent of the proposal. That's reassuring, but we need more reassurance than that. It is reasonable for citizens to expect rigorous analysis and comparison. Although the capacity mechanism proposed by the ESB for eastern Australia has elements similar to the capacity market currently operating in Western Australia, the ESB doesn't discuss the latter's cost and effectiveness. If one divides the WA total capacity payments by the amount of electricity through the South West Interconnected System in 2021, one gets a cost of about $35 per MWh. That is not a small number. It would add around half – in some states more – to average wholesale electricity costs in the NEM as they were in 2021. Is there some countervailing reduction in energy costs resulting from encouragement of more power into the system? There may be, but the case has not been made. A high proportion of payments are to the coal generators at Collie, mostly state-owned. When Premier Mark McGowan in June 2022 announced the closure by 2029 of these generators, he said this was necessary to avoid increases in power costs. Reliability of supply would be secured by new investment in wind and solar and state support for battery and pumped hydroelectric storage. The expensive capacity market was not mentioned as being relevant to reliability in this compressed transition.

The capacity mechanism proposed by the ESB could damage the one part of the NEM that had been working well before the Russian war crisis: the energy market. The competitive energy market has facilitated large investment in new and old technologies while adjusting to huge variations in economic and policy circumstances. Until the Russian invasion of Ukraine, it could be said to have delivered reasonably low wholesale power prices. It has demonstrated that periods of higher price encourage new investment and supply to bring prices down again. The aftermath of closure of the Northern power station in South Australia and Hazelwood in Victoria makes the point.

The energy market would be changed radically by a capacity market in the form proposed by the ESB. It would add new layers of administrative complexity. The consequences are not easily foreseen. Uncertainty about the effects of the capacity mechanism would discourage investment in renewable energy and storage. There are alternative approaches that are more straightforward.

Bruce Mountain et al. at Victoria University have suggested a Renewable Electricity Storage Target, with a similar structure to the RET. It accepts AEMO's assessment that $90 billion of private investment in storage is required by 2050. The authorities would designate a target level of storage corresponding to that number. Unlike the capacity mechanism, the incentive would cover decentralised as well as grid-level storage – important given AEMO's assessment that more than half of requirements would come from small batteries. Retailers and large users of power would be required to purchase and surrender storage certificates as they use power – just as they do with renewable energy in the RET. Market exchange would deliver the lowest-cost combination of storage. Mountain et al. see their proposal as an alternative to the Commonwealth's new Rewiring the Nation program, which supports state government investment in transmission to link Renewable Energy Zones to load centres. That is unnecessary. The Superpower needs much more storage and large transmission investment as well. The Mountain proposal could be a supplement rather than an alternative to investment in transmission.

I see merit in an alternative proposal by Tim Nelson and Joel Gilmore at Iberdrola for a reserve mechanism to balance the market whenever supply falls or demand rises to extreme or unexpected levels. Within the extremes, the energy market would be left to allocate resources and set prices. This has the great benefit of relying on an energy market that has been shown to work, except at clearly defined extremes where over-riding public benefits from reliability warrant state intervention.

My own view is that a variation on the Nelson–Gilmour approach would be most likely to secure high reliability at low cost. A Commonwealth-owned Energy Reserve would be given the task of providing enough capacity to ensure that supply always meets demand at a specified relatively high price, and that demand is always available to clear the market when negative prices reach high levels. This contrasts with current maximum and minimum price caps that leave excess demand when prices are high and excess supply when prices are low. Between the protected maximum and minimum prices, the competitive energy market would have full scope to invest in storage, peaking assets and demand management – free of the policy uncertainties that have inhibited investment in recent years. Most importantly, private investment would be free of uncertainties associated with a new capacity mechanism. The market would supply large amounts of decentralised storage in household systems and electric vehicles. It would encourage flexible use of power. Hydrogen electrolysers can be expected to emerge as the dominant users of electricity and would draw flexibly on huge amounts of power when prices were low and curtail demand when prices were high. The Energy Reserve would be required to intervene less and less as the Superpower developed – but would play a crucial role in ensuring reliability when required by unexpected extreme events.

Eventually, the Sungrid described in Appendix 1.2 would support the Energy Reserve, by providing large capacity reliably to supply power at low cost when required in any of the regions, and to absorb surplus power when it had little or no value in any region. Unexpectedly large deficits or surpluses of power are unlikely to emerge in all nodes of the Sungrid at the same time.

Appendix 1.1 describes in more detail how the Energy Reserve would work.

Secondary goal: The transmission network

The privatisation and corporatisation of transmission and distribution assets have been problematic. These activities are natural monopolies. Private ownership requires public regulation of investment and pricing decisions, as there can be no competitive market discipline and guidance. This regulation is notoriously difficult to do well. Australia has not done it well. We have moved from having among the lowest-cost transmission and distribution systems in the developed world to among the highest. We have invested the better part of a hundred billion dollars in expanding system capacities in the sixteen years in which the current rules have applied, but almost nothing to augment transmission to facilitate new renewables or to expand interstate interconnection.

Australia can correct that weakness over the next few years. AEMO has a major focus on transmission in its Integrated System Plan. Each of the states has developed its own program for augmentation of transmission, especially on deeper interconnection among the mainland states. The NSW government has developed a program linking renewable energy zones to the major Newcastle–Sydney–Wollongong region, supported by well-designed policies. Victoria has emphasised the deepening of interconnection with the better solar and wind resources of the west, while using the dense transmission system radiating from the old lignite generators in the Latrobe Valley to utilise wind offshore in the southeast. The Queensland government is working on a plan to transform transmission in the renewables-based industrial centres of central and north Queensland and unlock large industrial opportunities. SA has the large advantage of its old, retired coal generation having been located in a region of high-quality wind and solar resources. Effective use of legacy transmission assets takes it a long way. It has focused on deepening interconnection to the eastern states and using more effectively the excellent wind resources of the Eyre Peninsula. Tasmania has focused on the Mariner link to Victoria, which will increase its role in selling wind firmed by hydroelectric power to the mainland at times when it has most value. Western Australia has the foundations for an effective southwest regional transmission system in the legacy of connections from the old coal city of Collie to excellent renewable energy resources to the north (including the rooftop surplus from Perth, on to the Midwest including Geraldton), east (the wheat belt and the eastern goldfields) and south (through to Albany). Rewiring the Nation can add momentum to these state efforts.

One can quibble about the detail, but the accumulation of effort will go a long way to loosening what has been a tight bottleneck to expansion of renewable energy. The need is urgent and must be met soon if the 82 per cent renewables aim and associated benefits for electricity prices are to remain within reach.

A note on transport

Electric vehicles can play a big role in stabilising the electricity system as the proportion of solar and wind grows. This is in addition to the immense role that electrification of transport must play in decarbonisation, and the contribution it can make to reducing household and business costs and increasing national energy security.

Australia currently lags way behind all other developed and many developing countries in uptake of electric vehicles. In 2021, electric vehicles represented about 2 per cent of new light vehicle sales in Australia, 4 per cent in the United States, 6 per cent in Canada, 10 per cent in the EU, 12 per cent in the UK and New Zealand, 15 per cent in China, 36 per cent in Japan, 65 per cent in Norway – and 9 per cent in the world as a whole.

The Australian position is anomalous. Our world-leading position in household solar creates advantages in low-cost fuel supply for electric vehicles. Together with the relatively low cost of grid-scale renewable energy, this would see Australia moving to the front ranks of the world if it had comparable incentives for using electric vehicles. The policies of the Albanese government together with supportive policies of most state governments take Australia from a lonely trailing position to within the range of other developed countries. Given Australia's inherent advantages in use of solar power, that is probably enough to rapidly transform the role of EVs in Australia.

The expanded use of EVs can either increase costs and damage reliability of electricity systems, or reduce costs and increase reliability. For a positive outcome, charging of batteries must occur when power prices are low and the grid has excess capacity. Pricing of use of the grid needs to encourage this outcome, by making power available at very low prices when power is cheap and the grid free from stress, and at much higher prices at other times. Flexible pricing, varying with market conditions, is easily introduced before the large-scale deployment of electric vehicles. It would be more difficult after owners of a large EV fleet had become accustomed to inflexible pricing arrangements. With sound price incentives in place, expanded use of electric vehicles will spread the costs of the grid over larger electricity volumes and significantly reduce the costs of transmission and distribution per unit of power. It will enhance the reliability of power supply, by absorbing power at times of surplus and injecting power into the grid at other times. The average household uses many times as much energy in the car as the house, so the stabilising effects of this on the electricity system would be large. On the other hand, failure to make this change would see demand for grid services and energy increasing at peak times. This would exacerbate the reliability challenge and increase the cost of grid services per unit of power by forcing investment to expand grid capacity.

Transmission for a renewable energy Superpower
The transmission upgrade just discussed matches AEMO's 'Step Change' scenario. It would not support AEMO's 'Hydrogen Superpower' scenario. Much less would it support the growth in electricity

demand in the renewable-energy Superpower discussed in this book. The Superpower also requires a larger, lower-cost (per unit of power), technologically different transmission system to be built alongside, and to complement, that of the Integrated System Plan.

We can say with confidence that Australian export of zero-carbon goods in anything like the quantities discussed in this chapter would be impossible through incremental expansion of our existing grid. But plan the grid differently, and the scale and climatic diversity of Australian renewable energy resources would deliver power competitively to the main locations of the Superpower, and at the same time lower power costs and enhance reliability of power supply to the established electricity regions.

In *Reset*, I described large augmentation of the Australian transmission system to provide a high-voltage, direct current (HVDC) backbone to support the Superpower. We would strengthen established regional transmission systems. Complementary to that, we would build a new direct current backbone for the Australian electricity system, which would connect at one or two nodes in each state, where there were high-quality renewable resources and opportunities for building industrial load. I called this backbone the Supergrid. *Reset* commented that while transmission was a natural monopoly generally provided more efficiently by the public sector, that was unlikely given the state of Australian policy then. That may have changed with the new government. The Albanese government's Rewiring the Nation Corporation is intended to invest in deepening transmission networks – strengthening the established AC transmission systems around the old electricity regions with inter-regional interconnection.

Rewiring the Nation is focused on filling weaknesses and gaps that have been identified in the early stages of growth of renewable energy supply in the established electricity systems. Much more is required to support the Superpower. Of greatest value is a high-capacity HVDC backbone to the Australian electricity system. It would join all the regional power systems to each other and to the emerging nodes of zero-carbon industrial development in all of the mainland states and territories: a Sungrid.

Appendix 1.2 describes the Sungrid.

Second primary objective: Rapid deployment of new industrial technologies

The technologies that will drive the building of the Superpower, at least over the next one or two decades, are close to being commercially viable now. Their application involves risk. Those who take the risk of early deployment of new technologies create knowledge that is valuable to others and to the whole society, whether they succeed commercially themselves or fail.

This was implicitly acknowledged in the old Coalition government's Technology Roadmap. The Roadmap identified five priority areas: clean hydrogen, electricity storage, carbon capture and storage, low-emissions steel and aluminium production, and soil carbon sequestration. The inclusion of hydrogen and steel and aluminium technologies that are 'clean' but not zero-emissions was a scar from the climate wars. But the list contains several of the largest Superpower opportunities.

One positive aspect of the former government's Technology Roadmap was its elevation of the role of hydrogen. At the time of the Glasgow conference, the Morrison government announced support for a scheme of certification of green hydrogen, as a basis for emergence of a voluntary market. The new government would be wise to continue with this initiative.

There is no good reason for excluding a range of other technologies: pyrolysis for zero-emissions bio-oil and biogas and negative emissions char; use of bio-oil and biochar in chemical manufactures; zero-emissions nitrogenous fertilisers and explosives; sequestration in growing plants; zero-emissions production of silicon and other metals; and any other that involves innovation in zero-emissions production.

The policies of the previous government in practice favoured continued use of fossil carbon and involved high levels of official and especially ministerial discretion without transparent and stable rules. What is required is general provision of fiscal support to innovation in extending the zero-emissions economy. Support would be available at standard rates if clear conditions were met. The conditions relate to innovation either at the global frontiers of development of a new technology, or early applications in Australia of innovations that have been developed elsewhere. The Australian Renewable Energy Agency has a sound track record in managing grant schemes of similar kinds, and would be in a good position to play this role in a general program of support for innovation.

Reset advocated a change in the base of Australian corporate income taxation from standard accounting profit to cashflow. This would favour innovation, and more generally companies that were investing in growth over companies enjoying the rents from assets accumulated in the past. The proposed taxation reform would be particularly important in accelerating growth in newly competitive industries – with the zero-emissions industries of the Superpower in the first rank. It could even be adopted as a trial on an 'opt in' basis for companies investing in the zero-emissions economy, pending decisions on its general application.

Is there a need actively to stop or to discourage the old industries of the fossil carbon economy, alongside support for the new? The more important step is to recognise the external costs of activities which generate carbon emissions. In the absence of a general carbon price, the new government's proposed use of the safeguard mechanism in the Emissions Reduction Fund is a step forward. There are pressures for exclusion of trade-exposed industries. Responses should recognise that most other developed countries now have much stronger measures in place than Australia to restrict emissions. Australian producers of gas, coal and other emissions-intensive goods are benefitting commercially from other countries' restrictions. These issues were discussed analytically in the original Climate Change Review.[5]

There is pressure from people wanting strong action to combat climate change to restrict investment or production in the coal and gas industries. The international community decided at Kyoto and has consistently confirmed since that each country is responsible for emissions within its own borders, and not for those generated by use of its energy and other materials in other countries. Emissions from domestic production, processing and transport of coal and gas fall within Australia's commitments to reduce greenhouse gas emissions.

Methane and other greenhouse gases released in mining, transporting or processing coal, gas and oil, described as 'fugitive emissions', are now the largest source of Australian emissions after electricity generation. They have been growing rapidly over the past two decades and are well above 2005 levels. If the gas and coal industries proceed with current plans to expand some mines and to open new ones, fugitive emissions will continue to rise, and the prospect of reducing emissions by 43 per cent by 2030 will diminish. The main coal and gas jurisdictions, especially Western Australia, Queensland and the Northern Territory, will breach their own emissions-reduction commitments.

Technologies for measuring emissions reliably and at low cost from satellite observations mean that any errors in reported data will be found out. It is much better for the reputation of Australian regulatory agencies and companies that Australians discover and correct errors before they become the subject of international reaction.

There is a strong case for leaving decisions on investment and export to private corporations that comply with Australian law and regulatory conditions, and for requiring the removal or offsetting of fugitive emissions. The financial burden of immediately removing net fugitive emissions would be manageable at this time of extraordinarily high prices and corporate profits. Every tonne of emissions in Australia from coal and LNG production imposes a cost on the international community and is attributed to Australia. Requirements of full offsetting of fugitive emissions would lead to larger efforts to reduce them. It would be reasonable to insist on full offsetting now. If phased in over the period to 2030, it would support Australia's acceptance of the Glasgow methane pledge, guarantee achievement of the 43 per cent emissions reduction target and underwrite strong outcomes for emissions reductions by 2035.

Policy on carbon credits from managing the landscape

We are at the early stages of unlocking the value for Australia and the world of using land to grow biomass for zero-emissions industry and generate carbon credits from absorbing carbon in plants and soils.

We have made a start over the past decade by establishing a system of accounting for increases in carbon in land and plants and rewarding increases with carbon credits ('ACCUs') that can be sold into voluntary markets, to the Commonwealth government's Emissions Reduction Fund (ERF), or to firms required to surrender permits to the Clean Energy Regulator for shortfalls within the fund's safeguards mechanism. The voluntary markets have grown rapidly over the past one and a half years. ACCU prices were rising strongly towards levels in other developed countries until a decree in early 2022 by the Minister for Emissions Reduction in the Coalition government, Angus Taylor, fundamentally changed contractual arrangements for delivering permits under the ERF and prices fell by half. They partially recovered the losses through the middle months of 2022.

There were a number of changes to land carbon methodologies through the life of the Coalition government that weakened the integrity of the system. In addition, time and experience revealed weaknesses in some methodologies. The former chair of the Commonwealth government's Domestic Offsets Integrity Committee, Professor Andrew Macintosh, offered a far-reaching critique of the ERF in early 2022, saying that a high proportion of ACCUs had been allocated for non-existent accretion of carbon stocks.[6]

Blair Comley, a former secretary of the Commonwealth Department of Climate Change, led the production of a paper on land carbon for the EY Net Zero Centre. He notes that credits from landscape carbon are centrally important to global achievement of net zero and to building the role of Australia in landscape carbon. He observes that for landscape carbon to play its full role, it is necessary to ensure rigorous accounting for change in the carbon stock. There have been loose practices in the past. From now on, it is essential that only genuine, permanent negative emissions be rewarded. There should be no role for offsets awarded for simply reducing emissions in some activity. Failure of rigour in accounting for carbon credits will lead to exclusion from the increasingly important global compliance and voluntary markets for carbon. This is a timely and well-argued caution. The new government responded appropriately to the EY and Macintosh critiques by announcing in June 2022 a review of the ERF chaired by a former Commonwealth Chief Scientist, Ian Chubb.

Isabelle Grant in Chapter 7 explains the case for reform of credits for increasing carbon in the landscape. Big steps need to be taken in two areas: accounting for changes in carbon stocks; and measurement technologies. We must move away from partial methodologies to comprehensive measurement of changes in carbon stocks from transparent, authoritatively defined and stable baselines. Participants would be rewarded for all increases in carbon stocks in soils or plants, whatever their origin. They would need to take responsibility and pay for any depletion of stocks.

Realisation of the potential contribution of landscape carbon to the global movement to net zero requires measurement technologies that can be operated on a large scale at low cost. This requires use of remote sensing from satellites or drones. Promising approaches have been identified, but they need to be tested against traditional methods and developed commercially. The Coalition government took a step in that direction by including soil carbon in its five technologies warranting support. That needs to be extended to carbon in plants and accelerated.

A new economic geography
The new development would be disproportionately in provincial and rural Australia. There can be early growth in new forms of employment on a locally transformative scale in niches of surplus capacity of transmission and other infrastructure. The Barcaldine Renewable Energy Zone industrial precinct, described in Chapter 8 by Susannah Powell, is a model of what is possible. Different sets of industries will be suited to the different resources of each location. There is commercial room for a dozen or so small-scale urea plants in country towns across Australia, each providing several hundred local jobs, increasing supply security for local agricultural and pastoral industries and lowering delivered costs of fertilisers. This could happen quickly.

The various dimensions of carbon farming could quickly make substantial contributions to incomes growth in all regions of rural Australia. The first iron metal production would occur commercially in small-scale projects in provincial towns with some history of industrial activity. It would draw on legacy electricity, water, port and other transport infrastructure. The early candidates would be close to iron ore mines in regions with excellent renewable energy resources – perhaps small deposits which were on the margin of profitability when supplying export markets. The Upper Spencer Gulf in South Australia, southwest and midwest Western Australia, North Queensland, Broken Hill and Newcastle in New South Wales, and northern Tasmania all have opportunities for early progress.

The big centres of the Superpower are likely to emerge in provincial cities with industrial traditions close to excellent renewable energy resources. Adjacent biomass resources are an advantage, and in Queensland's Rhombus of Reliability will be a distinguishing feature. They will require new transmission and other infrastructure on a massive scale.

We can expect to see the emergence of at least one or two major centres of new economic activity in each state and the Northern Territory.

The Upper Spencer Gulf is prepared by renewable energy and mineral resources, history and support from the SA government to be an early leader.

The emergence of a major industrial hub in Dubbo and the central west of New South Wales, and a great expansion of activity in Newcastle, are supported by the NSW government's well-designed renewable energy program.

Tasmania has large advantages in renewable energy supply. Excellent wind resources can be supported by the shifting of hydroelectric operations to an explicitly firming role. Firming can be extended by utilising the island's immense pumped hydroelectric storage resources when minerals processing and other green manufacturing have exhausted the peaking potential of traditional hydroelectric power generation. Ports near established industrial towns allow local mineral resources (iron ore, some non-ferrous metals) to be augmented by materials from other states – as they have been for many generations with manganese ores from the Northern Territory and non-ferrous metal ores from New South Wales via Port Pirie in South Australia.

Queensland has excellent prospects in the Rhombus of Reliability. The Rhombus is bounded by the industrial ports of Gladstone in central and Townsville in north Queensland. It has water and biomass resources that are exceptional in Australia and potential for firming of renewables with pumped and conventional hydroelectric power. It extends west of the Great Dividing Range to world-class solar resources in Hughenden and Barcaldine, and high-quality wind resources near both. The ports built for coal near Mackay are well placed to utilise potential for bio-industry. The coalmines through the Bowen and now extending into the Galilee Basin leave a legacy of electricity transmission, transport and water infrastructure that can support medium-scale manufacturing. The northwest minerals province centred on Mount Isa and joined by railway to Townsville provides a wide range of mineral ores for processing. Pockets of resources of energy transition minerals as well as iron can diversify the manufacturing base. Some of the abundant bauxite resources of the Gulf of Carpentaria are currently converted into alumina and aluminium in Gladstone and the scale of refining and smelting can expand.

The Northern Territory potential is underdeveloped but considerable. The solar resource south of the coastal summer cyclones, rain and cloud will support the Sun Cable project and a transmission backbone around which the flesh of a large-scale domestic power system can be built. Wind resources are of high quality on higher country to the south and southeast, including in the vicinity of Alice Springs. As with the Rhombus of Reliability in Queensland, there is large opportunity for sustainably growing biomass outside the arid zones. The Northern Territory is richly endowed with mineral resources: aluminium ores mined and previously processed into alumina at Nhulunbuy; manganese in the Gulf of Carpentaria currently processed in Tasmania and abroad; iron ore; and energy transition minerals in many locations. In the Middle Arm industrial precinct, the immediate allure of gas has diverted attention from sustainable opportunities from renewable energy but can help to establish the infrastructure to support it.

In Western Australia, the southwest region based in the old coal and power generation of Collie and the midwest around the port of Geraldton both have access to excellent wind and solar resources.

Collie can utilise the transmission lines that once carried coal power north to Perth and Geraldton, east through the wheatbelt to the goldfields, and to Albany and the Great Southern. Well-established infrastructure and the availability of a wide range of industrial skills will support early industrial development on a substantial scale in the southwest and then the midwest. The full development of the Pilbara will be the apotheosis of the Australian renewable-energy Superpower. One day, the incomparably reliable and predictable energy in what are close to the world's highest tides will complement the solar away from the coast, the wind from behind the eighty-mile beach between Port Hedland and Broome, and the pumped hydro storage in depleted mine sites. The Pilbara will eventually be an immense centre of minerals processing, with iron smelting at its core, and perhaps of ammonia exports.

Victoria has large opportunities around the established industrial centres of the Latrobe Valley and Portland. Both have access to considerable biomass resources which, in combination with the state's commercial and manufacturing industrial strengths, have the potential to make Victoria a major global centre for zero-emissions chemical industry. The legacy of transmission from the coal days supports supply of renewable electricity from favourable locations. There are pockets of mineral resources for processing, but large-scale development is likely to depend on shipping ores from other states – as the Portland aluminium smelter now draws alumina from Western Australia. Victoria's largest role in the Superpower is likely to be as a centre of finance, management and innovation for the new economy. It is well placed to capture a high proportion of the advanced manufacturing opportunities that emerge.

Building the Superpower will transform the economic geography of Australia. The main metropolitan areas have advantages for niches of advanced manufacturing in which global competitiveness is less sensitive to high costs of transmission from the renewable energy zones. The cities will benefit symbiotically from the new dynamism as suppliers of specialised services and some goods, and their residents will play many leadership roles. But they will not be the centres of growth. Provincial and regional Australia will see faster growth than the major cities, for the first time on a sustained basis since Federation. The old dream of decentralised economic growth will emerge simply as the unfolding of a new pattern of economic development. The early movement will be most rapid in well-established old industrial centres. New dynamism will emerge in small towns which can utilise social and economic infrastructure installed to serve larger populations than are present today.

Indigenous Australians will have an enhanced role in Australia's new economic geography.

The new economic geography will reshape the political map. The Liberal National Party in Queensland and the National Party elsewhere held its electoral ground in 2022 against the losses of their Liberal partners. To continue to hold that ground they will need to reflect the new economic base of rural and provincial Australia, built on rural Australia's participation in the decarbonisation of the Australian and global economies.

Reset

The electoral flood has changed what is possible in Australian policy, and above all on climate and the Superpower Transformation. The traps left from the climate wars have been sprung by the Russian invasion of Ukraine, and the government is struggling with the consequences. It is constrained by promises made before the 2022 election that it would not do much to disturb settled ways.

Australia's position as a large exporter of coal and gas gave its government the choice of protecting the living standards of ordinary people from the Russian energy price trap – or leaving undisturbed the windfall profits from the Russian war. Protecting living standards of Australians would come at the cost of a minority proportion of the windfall profits from the war – disappointing producers who are unaccustomed to disappointment and react badly to it. It is not yet clear in mid-2022 how the government will exercise that choice. A decision to let the high global energy prices from the Russian war reduce real wages and living standards for most Australians would test domestic political cohesion, as high energy prices are testing the internal cohesion of Europe and the United States. Such a decision would actually increase the profitability of established renewable energy generation, and to some extent encourage investment in the new economy. However, the encouragement would be tempered by expectation that high electricity and gas prices would remain in place for too few years materially to affect new investment in power generation.

The new government faces a separate longer-term choice between embracing the energy transition or retreating to old energy strengths in response to the Russian energy crisis. The early signs are that it has chosen the Superpower future, with all of its opportunity for Australia and the world.

While the government in the election campaign promised not to do much to solve many of the deep economic problems left from the Dog Days, on climate and energy it left itself room to make a good start in new directions. The new parliament is ready to support strong shifts towards zero emissions and the building of Australia as a renewable-energy Superpower, and to back the government on going faster and further in second and third terms. What will we make of this opportunity? Let's recall what is at stake.

Suggestions made in this essay would put us on a path to meet our new commitment to the UN to reduce global emissions by 43 per cent by 2030. We would be seen to be on a path to reducing emissions by

75 per cent from 2005 levels by 2035 and zero net emissions in the 2040s. They would allow us credibly to commit to the pledge of all other developed and many developing countries to reduce methane emissions by 30 per cent by 2030. They would allow us credibly to join the developed country commitment to phase out coal for power generation by 2035.

These developments together would transform our position from the developed world's biggest drag on progress on climate change to a positive force. That change would protect our access to developed country markets for high-value food and other exports that would otherwise become more and more difficult from 2025. It would give us access to rapidly expanding markets for zero-carbon goods in which we have comparative advantage, and for expanding quantities of carbon credits from rural Australia. We would still have to be deft in diplomacy and in management of domestic carbon markets of various kinds, but there would be a path to success if we managed things well.

Australian exports of zero-carbon goods and services would soon be seen as a substantial boon to the global fight against climate change. Our export of green hydrogen and ammonia would be seen as a substantial contribution to Europe and Northeast Asia managing the disruption of Russian exports of coal, oil and gas without for long going backwards on decarbonisation. The industrial economies of Europe and Northeast Asia would be realising that the path to decarbonisation of steel, aluminium, silicon and many other energy transition minerals is through greater use of zero-emissions processed materials from Australia. Singapore would be a step closer to using electricity transported by submarine cable from Australia, and to demonstrating the value of that trade to other Southeast Asian countries. Negative emissions from the Australian land sector would be recognised with increasing clarity as one of the legitimate means of offsetting emissions that were difficult to remove at home.

Australian supplies of silicon, lithium and many other processed energy transition minerals would be relieving anxieties about materials being available to support decarbonisation in the rest of the world. Economically fruitful reorganisation of the Australian economy over the next several decades would make a large direct contribution to decarbonisation of the global economy. Our own movement to zero would reduce global emissions by about one and a quarter per cent. The full Superpower export of zero-carbon goods and carbon credits would reduce them by about 7 per cent. The immense investment required to achieve this outcome would mean that we would not get to this endpoint until after the middle of the century – but the last stages would be helpful to Australia's neighbours in developing Asia reaching their own zero-emissions goals. That means Australia could eventually contribute directly over 8 per cent of the reduction in global emissions. This is much more than continental Europe (including the UK) removing all emissions. It is much more than Japan reaching net zero twice.

Successful transition would allow the achievement of full employment in Australia with rising living standards for the general population and a reasonable amount of foreign debt. This is a marked change in circumstances, as we would otherwise be mired in stagnation with our traditional exports blocked in various ways. The total investment is daunting – but cool assessment would show that the private sector had done as much as this before. The total investment required as a share of the economy from now to the 2050s would be similar to the share of the economy allocated to investment in mining to meet the demands of the China resources boom.

The restructuring of the economy to build the Superpower would require the reset of many attitudes and policy. We would need to invest heavily in upgrading education and training for old and new Australians. A disciplined immigration program would need to focus strongly on valuable skills. Labour would be scarce and valuable, and jobs would not. We would need to be ready to let firms and industries die if they could not support high and rising Australian living standards.

We would need renewed respect for knowledge and analysis in the policymaking process. We would need to learn again the disciplines of governing in the national interest and resisting pressures from vested interests.

We face an immense opportunity. Our decision to accept the opportunity or to turn our backs on it is at least as important for Australia and vastly more consequential for the world as a whole as the choices Australia made on the two previous occasions when it successfully chose great opportunity requiring major structural change: in postwar reconstruction, and in the reform era of the late-twentieth century. The Albanese government's commitment to a White Paper on full employment in its first term of office, redolent of postwar reconstruction, provides an appropriate context for considering and taking national decisions of historic dimension.

More Australians, year by year, have been moving onto the bridge to the Superpower. The electoral flood of 2022 clears the path to the Superpower Transformation.

Appendix 1.1 The Energy Reserve

The Energy Reserve would be a Commonwealth entity, with the task of achieving high degrees of reliability in electricity supply at the lowest possible cost. It would be charged with ensuring that supply of power would match demand in all regions under almost all conditions. The tolerance limits of reliability would be specified in the Energy Reserve's governing rules.

The government would specify the range within which market exchange determines prices and investment in storage and other assets to balance the variability of wind and solar power. The range would be wide enough for large interventions by the Energy Reserve to be infrequent. The price limits would be informed by analysis. The wider the range, the greater the private incentive to invest in storage and peaking assets and to manage demand. The wider the range, the lower the required public investment in reliability. My initial suggestion would be a lower limit at levels now operating in the NEM (minus $1000 per MWh). It could be higher (less negative). My initial suggestion is for an upper limit of $5000. This is considerably lower than the current upper limit in the NEM ($15,500), but low enough for it rarely to be exceeded over lengthy periods. It could be some lower number, say $3000 – at a cost to incentives for private investment in storage and peaking assets. I use a minimum of minus $1000 and maximum of $5000 in the example below without presuming that the authorities would settle on these numbers.

The Energy Reserve would control enough storage capacity to hold the price at minus $1000 whenever the market fell to that level. Judicious use of privately owned electrolysers producing hydrogen would ensure that negative prices were rare and seldom, if ever, near the specified minimum.

The Energy Reserve would hold enough generation and storage capacity to keep the price at $5000 whenever the market took it to that level. To provide certainty about the behaviour of a major state participant in the market, the Energy Reserve would not sell power when the price was below $5000.

To be in a strong position to protect the upper limit on prices, the Energy Reserve would seek to have its storage full whenever prices were approaching the limit. This would often require purchase of power at prices above the minimum. The Reserve would set clear rules under which it would purchase power.

The Energy Reserve would be instructed to meet its objectives at the lowest possible cost. It would generate revenue from arbitrage – purchasing power for storage when prices were low and selling it at $5000 per MWh. The arbitrage revenue would be offset against cost of storage. Storage assets would therefore support higher costs per unit of power and energy than thermal peakers that use expensive fuel and lack opportunities for arbitrage. The Energy Reserve would need to draw on authoritative assessment of future demand and supply and how they might vary with economic, climatic and other conditions within each region. Such assessment is now the responsibility of AEMO. It does the job well and can provide the market perspectives to guide the Energy Reserve.

The Energy Reserve would have to ensure that it can cover both short and long periods of release of power to protect a $5000 maximum price. This would require a mixture of storage and peaking assets.

Batteries could cover shortfalls over several hours – or longer as greater use of flow and other long-duration batteries leads to lower prices. They could be backed up by gas peaking generators. Longer shortfalls could be covered by long-duration pumped hydro and gas peaking assets, and possibly mothballed coal generation assets maintained and held in reserve for rare events.

The Energy Reserve would purchase rights to use electricity peaking generation and storage wherever it could find them, within its mandate to achieve a specified degree of reliability at the lowest possible cost. Expected arbitrage earnings would be taken into account when comparing costs.

Who would own the storage and peaking assets? The Energy Reserve would have access to the Commonwealth's balance sheet, so it would be a low-cost owner of capital-intensive assets. Its participation in ownership could facilitate rapid development of some storage and peaking assets. However, there are generally efficiency advantages in private ownership. There are other advantages in leasing. Costs might be reduced by the Energy Reserve leasing part of the capacity of a privately owned peaker or storage asset, and leaving the private owner to participate in market exchange with the balance of the capacity. Alternatively, the Energy Reserve could own the asset and lease surplus requirements for private use within the range of prices in which the market allocates resources. Most of the private value from long-duration storage assets like pumped hydro are from the early hours of capacity, so the Energy reserve may be able to purchase the later hours at relatively low cost. Sharing the use of an asset by lease or shared ownership could also be over time. For example, the Energy Reserve could lease the use of an asset for a number of years, with use rights reverting to a private owner after that time. This might be valuable, as demands on firming may fall after the completion of the transition to zero emissions.

The 'technology neutrality' mantra from the culture wars should be dropped in favour of an environmentally and economically rational approach. Calculation of the costs of peaking services would embody the cost of carbon emissions. Suppliers of services could be required to purchase and surrender credible carbon offsets, including for services supplied by the Energy Reserve itself. With this recognition of the cost of carbon emissions, the Energy Reserve could draw power supplies from their lowest cost sources.

Would demand management be part of the Energy Reserve's brief? Industrial and other processes which can reduce demand at little cost when prices are high and expand use of power when prices are low will become much more important in the energy system. Production of hydrogen by electrolysis can be varied within a wide range and will eventually absorb a high proportion – half and much more – of electricity output. The cost of hydrogen storage is low compared with requirements to purchase electricity at small premiums above average levels, so electrolysers will be fully engaged in their own arbitrage within the price range protected by the Energy Reserve. Private demand management will expand within the energy market. This development alone will greatly reduce the frequency and extent of intervention by the Energy Reserve. My own inclination is that demand management is better left to private markets.

I have suggested in *Reset* that one way of beginning the building of a reliability reserve would be to separate Snowy Hydro's peaking and storage assets from its conventional hydroelectric and retail businesses. The Energy Reserve could be responsible for managing established gas generation and pumped hydro storage and also for completion and operation of Snowy 2.0 and the Kurri Kurri gas and hydrogen generation facility. The knowledge that these assets in the hands of a sovereign entity will enter the market in ways not known to other participants has been a substantial deterrent to private investment in peaking and storage assets. Snowy pumped hydro storage and gas peaking (with off- setting of emissions) would provide great depth of storage. The Energy Reserve may choose to lease into private markets the early hours of Snowy pumped hydro storage and gas-hydrogen peaking, for use within the range of prices within which market exchange occurs, and maintain control of long-term storage.

The greatest demands on the Energy Reserve would be over the next dozen years of transition to zero net emissions from electricity generation. Demands would be heavy soon after each retirement of a coal generator – the more so when two or more closures occur close in time. Private markets for arbitrage will become more sophisticated and absorb more of the balancing load over time. Decentralised storage in home batteries and electric vehicles will add immense depth to private storage which will help to keep prices within the specified range. The growth of the Superpower economy with large-scale electrolysis to produce hydrogen for industry and export will greatly expand the size of the system and the proportion of demand absorbed into highly flexible uses. The Superpower economy will expand inter-regional interconnection, allowing diversification of solar and wind resources and the absorption of regional shocks over a larger market. The Sungrid described in Appendix 1.2 would greatly reduce the costs of providing reliability in the established regional electricity systems.

A new approach to reliability is needed now. The ESB says that it will take until 2025 to implement its capacity mechanism. The Energy Reserve could be established immediately, based on AEMO's Reliability and Emergency Reserve Trader, strengthened by addition of the Snowy assets (now wholly owned by the Commonwealth and so available to quick decisions) and built quickly from that foundation.

The Energy Reserve conception of a capacity mechanism would have the great advantages of being focused on a single clear objective: ensuring that supply matches demand at any price. It would have the great advantage of simplicity, performing its function without tangling with transactions inside the energy market. It would have the great advantage of having the energy system harnessing the power of market exchange in all but extreme circumstances.

Appendix 1.2 The Sungrid

I have worked with London-based consultancy HVDC TECH on a high-level description of a multi-terminal complex of electricity super-highways as an overlay to the existing power systems of the NEM and the three other substantial mainland systems: the South West Interconnected System, the Pilbara region (undergoing internal interconnection, including eventually to the Asian Renewable Energy Hub) and the Darwin–Katherine system (to be connected to the emerging Sun Cable system). Let's call it the Sungrid. The Sungrid would have seven nodes: two in Western Australia and one in each of the other mainland states and territories (with the ACT connected through New South Wales). The proposed routes follow established road and rail corridors to minimise incremental environmental and social disruption and facilitate construction logistics.

The system would be built with high-voltage, direct current (HVDC) technology. Use of direct current reduces transmission losses over long distances. HVDC has the disadvantage that it is very expensive to introduce additional terminals for taking in and discharging power. Each of the nodes in our proposed Sungrid was selected for rich local renewable energy generation opportunity and connection to the established regional systems and location close to major loads and potential for developing local industrial loads. Optimisation of the system before building the Australian Sungrid would involve testing the suggested nodes against alternatives.

The following nodes were used as a basis for costing the system:

1. Dubbo in the Central West–Orana Renewable Energy Zone in New South Wales. This is potentially an important regional industrial centre and is to be linked to Newcastle and the dense metropolitan power system of New South Wales with its HVAC transmission connections.

2. Mildura–Red Cliffs in northwest Victoria. This has the richest

solar resource in Victoria and is not far from the excellent wind resources of western Victoria. It is adjacent to a terminal in the NSW–SA interconnector that is under construction. It has established AC connections to the mining and industrial centre of Broken Hill, which has opportunities for large expansion based on iron, base metal and other mineral resources. It would need to be connected into the dense Victorian regional system by double 500kv AC lines in an inverted Y, south to Horsham and on to Heywood–Portland and to North Ballarat – Mooroobool. Optimisation of the Sungrid would involve detailed analysis of alternative locations for the Victorian node, including at Kerang with connection to the proposed new Victoria–New South Wales interconnector.

3. Port Augusta (or rather, the nearby Davenport network centre) in South Australia. This was the centre of coal-based power generation in South Australia and has thick transmission connections to the industrial cities of the Upper Spencer Gulf (Whyalla and Port Pirie) and south to Adelaide. HVAC transmission links are being strengthened to the superb wind resources of the Eyre Peninsula. The Upper Spencer Gulf with its adjacent mineral resources and industrial history is in a good position to emerge as an early leader of renewables- based new industrial activity.

4. Northam, about 100 kilometres east of Perth. Strong connections would be made west to Perth and the industrial precinct of Kwinana, south to Collie (the old centre of coal power generation and well located for a wide range of minerals processing and other industrial activities) and north to Geraldton (with excellent wind and solar resources and large opportunities for minerals processing). Each of these would emerge as a major centre of new industrial activity.

5. One point in the Pilbara, which would become the node for an integrated regional electricity system across the region. Mount Tom Price is suggested as the initial place for assessment, as the massive, depleted mine pit holds opportunities for pumped hydro storage that would have national as well as regional significance. The Pilbara is the location of the proposed Asian Renewable Energy Hub, which is well placed to be a massive exporter of hydrogen and ammonia. The Pilbara region has excellent solar resources away from the cyclonic coast, and also wind. It has immense opportunities for iron and other minerals processing. It is the most favoured location in the world for the largest of the Superpower industries: green iron. Its disadvantage is a high-cost structure based on highly profitable iron and gas mining, so that it will share iron processing of Pilbara ores with other Australian industrial ports. Others will be more competitive in going beyond iron smelting into steelmaking and the production of some downstream steel products.

6. Alice Springs is the centre of rich solar and wind resources. It is in a highly mineralised region with many processing opportunities. It would be the point of connection with the Darwin–Katherine electricity system and with the immense Sun Cable project for export of solar energy to Singapore and potentially to Indonesia and the Asian mainland. There would be large opportunities for productive interaction with the Sun Cable project, releasing power from the Sungrid for export by Sun Cable when it is abundant and cheap, and absorbing power from Sun Cable when it is available in larger quantities than required for export. There are large opportunities for minerals and biomass processing at industrial precincts at Middle Arm, Tennant Creek and,

with Indigenous support, the aluminium mining and processing town of Nhulunbuy.

7. Barcaldine, in the central west of Queensland. It has superb solar resources and is adjacent to the excellent wind resources of the western edge of the Great Dividing Range. It would be the point of connection through high-voltage HVAC cables with the huge industrial potential of what is described as the Rhombus of Reliability – encompassing the established industrial and commercial port cities of Gladstone, Rockhampton, Mackay and Townsville, the legacy infrastructure from coalmining in the Bowen and Galilee Basins and immensely valuable water and biomass resources. The Rhombus would connect at Hughenden with the proposed copper string transmission line from Townsville to the rich minerals region centred on Mount Isa.

Our cost estimates are based on a uniform 4 GW capacity DC line joining all nodes. They include costs of terminals at each node for connecting DC to AC systems. Costings were based on those in similar conditions in the west of the United States at an exchange rate of one Australian dollar to 70 US cents. They exclude costs of interest during construction and Australian taxes. Optimisation would examine the merits of increasing the capacity of some or all links.

There are several opportunities for large additions of power to the Sungrid from new sources that may have some role to play in providing low-cost reliable electricity in the Superpower.

Fortescue Future Industries has taken a close interest in the massive hydroelectric potential in the south-flowing rivers on the island of New Guinea. One of several sites on the Purari River, Wabo, has been the subject of complete and favourable feasibility studies in the past – in the last years of colonial administration extending into early independence and again a dozen years ago by a joint venture of Origin Energy and Papua

New Guinea Sustainable Development Limited. If FFI or others were to generate large amounts of electricity from these resources, the low-cost points of entry into the Australian electricity system would be into Sungrid at Barcaldine, or through the mining and minerals processing town of Nhulunbuy, in east Arnhem Land, if Indigenous communities saw value in it.

Reset mentioned the possible advantages of drawing deep peaking services from gas generation at Moomba in the northeast of South Australia, with storage of captured carbon dioxide in the depleted reservoirs of the gas fields. I noted that this may be one of the few examples of suitable geological structures adjacent to fossil energy resources with potential for cost-effective production of electricity with net zero emissions. This would require HVDC transmission to the Sungrid at Port Augusta (about 700 km) and possibly Barcaldine (about 950 km) as well. Other possible sources of peaking power from gas, using established gas generators rendered redundant by high gas prices, and with sequestration in adjacent depleted gas structures, are in Middle Arm in the Northern Territory (connecting to Sungrid through the Sun Cable network and the Alice Springs node) and through the Pilbara (connecting at Mount Tom Price or the alternative Pilbara node).

There was considerable discussion on the Coalition side of Australian politics after the May 2022 election of a possible role for nuclear energy in a net-zero-emissions electricity system. Analysis one and a half decades earlier had suggested that some Australian nuclear electricity may have economic value later in the process of decarbonisation, from the late 2030s. Since then, the prodigious reduction in costs of renewable energy and battery storage have made it unlikely that there would be a place for nuclear power in an economically efficient Australian electricity system. Apart from its high costs, disadvantages include extreme inflexibility of output, when flexibility is required to balance variable output of solar and wind power. However, should changing circumstances open a place for nuclear power, the Sungrid would create a suitable location for it in stable geology adjacent to uranium mining near the intersection of the Eyre and Stuart Highways in South Australia.

The New Guinea, Moomba or other gas and nuclear options would be available should large new additions of flexible (hydroelectric or gas peaking) or baseload (nuclear) power be helpful in lowering costs for reliable operation of the Superpower electricity system.

The Sungrid would greatly reduce the need for peaking power and storage in all the main current and prospective load systems in Australia. For the set of Superpower industries hanging off each node, it would be a low-cost source of arbitrage. It would absorb power when it was available locally in abundance, and supply it when it was locally in short demand. Connected by upgraded regional transmission systems into the respective metropolitan markets, this would effectively integrate the four mainland regional systems in the NEM with each other and with the three regional power systems outside the NEM. The east–west dimension would extend solar availability by an average of 2.5 hours each day. Eastern Australia would supply the west coast through the latter's morning peak and western Australia would supply the east coast through the latter's evening peak. The north–south dimension would extend southern solar availability by half an hour on a midwinter's morning and an hour on a midwinter's evening (comparing sunrise and sunset, Tennant Creek and Melbourne). It would greatly improve the quality and reliability of solar power to southern Australia through the winter. There would be transformational diversification of wind resources.

The grid would be built in stages, which could follow each other or overlap.

Stage 1 would be Barcaldine–Dubbo–Mildura–Port Augusta (distances taken from existing road and rail corridors):

- Barcaldine–Dubbo. 1200 km; 4.8 per cent line losses.
- Dubbo–Mildura (Red Cliffs). 770 km; 3.5 per cent losses.
- Mildura–Port Augusta. 500 km; 2.6 per cent line losses.
- Total stage 1: 2470 km; $13.9 billion capital expenditure.

Stage 2 would complete the eastern pentagon of the Sungrid with Port Augusta–Alice Springs–Barcaldine:

- Port Augusta–Alice Springs. 1250 km; 5 per cent line losses.
- Alice-Springs–Barcaldine. 1400 km; 5.6 per cent line losses.
- Total stage 2: 2650 km; $12.8 billion capital expenditure.

Stage 3 would complete the Sungrid by filling out the western quadrilateral of the Sungrid with Port Augusta–Northham–Pilbara–Alice Springs.

Figure 1.1 The Sungrid.

- Port Augusta–Northam. 2200 km; 8.2 per cent line losses.
- Northam–Pilbara. 1100 km; 4.4 per cent line losses.
- Pilbara – Alice Springs. 1600 km; 6.4 per cent line losses.
- Total stage 3: 4900 km; $22.4 billion capital expenditure.
- Total Sungrid: 10,020 km; $49.4 billion.

Figure 1.1 shows the location of the Sungrid in stylised form, together with AC connections that would need to be strengthened in Queensland, New South Wales, Victoria, Western Australia and the Northern Territory.

We need to start transforming the transmission system now. The established momentum of the states in strengthening transmission to allow greater connection of solar and wind power is augmented by Rewiring the Nation. The building of the Sungrid could be a commitment at the federal election due in 2025, for implementation over the subsequent decade. That would set Australia up for reliability and low cost zero-emissions electricity by the mid-2030s, to support building the Superpower and ensuring high reliability and low cost in the established metropolitan centres of large-scale power use.

2

THE DIMINISHING CARBON BUDGET AND AUSTRALIA'S CONTRIBUTION TO LIMIT CLIMATE CHANGE

Malte Meinshausen, Zebedee Nicholls, Rebecca Burdon, Jared Lewis

Where does the world stand?

Throughout human history, great transformations have often begun through tiny, almost imperceptible changes accumulating until suddenly a big shift happens almost overnight. That gives hope for climate mitigation action.

Many climate and emissions indicators point in the wrong direction. Global greenhouse gas emissions have steadily increased in the thirty years since the Rio Convention in 1992. We are already observing the devastating effects of unprecedented floods, an increase in the size and intensity of bushfires beyond anything previously recorded, and repeated episodes of coral bleaching that trash the hopes of preserving the Great Barrier Reef for future generations.

And yet, despite this bleak picture, there are indicators which give greater cause for optimism than in previous decades. Achieving the sudden, large-scale societal shift required to avoid the most devastating consequences of climate change now seems more possible than before. This is mostly due to the changing economics of renewable power. Plummeting renewable energy costs[1] are leading to emissions reductions, even in countries without effective climate policies. That transformational development was driven by early and costly policy initiatives largely motivated by an understanding that rapid decarbonisation was necessary to avoid dangerous climate change. The high feed-in tariffs for solar electricity in Germany are one of the key measures that spawned a global solar photovoltaic (PV) industry.[2]

Electrification and renewable energy are now the least expensive way to provide power to most of the world, including its poorest areas. 'Doing the economic thing' and 'doing the right thing' are now one and the same. This has happened faster than many had predicted. Installed solar PV around the world strongly surpassed projections on the basis of current policies, as shown in Figure 2.1 – which reveals as much about the success of solar PV as it does about the hesitancy (or inability) of modellers to project transformative shifts in markets.

However, fossil fuels still sit at the heart of many economies. This has been highlighted by Europe's struggle to wean itself from Russian oil and gas in response to Russia's invasion of Ukraine. Will Russia's war accelerate the transition to renewable energies? Warlike mobilisation of energy efficiency measures, heat pumps and renewable energy deployment may occur in the gas-dependent areas of Europe. In other regions, the drive for energy independence may stimulate investment in domestic oil, gas and coal supply. The drive for energy independence is supercharging investment in whatever energy is available from domestic sources. The

Figure 2.1 The actual development of solar PV additions per year (solid black line) according to the IEA renewable energy statistics – compared to gross additions under the 'stated policies scenario' of successive annual World Energy Outlooks (grey dashed lines), the forecasts in the IEA Renewables 2021 report (grey shaded area) and those of Bloomberg New Energy Finance (dotted line).

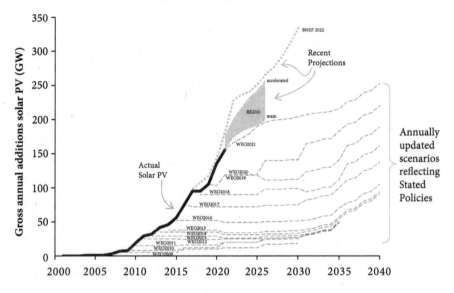

Source: Compiled by Dr Simon Evans, CarbonBrief, updated from 'Exceptional new normal' Carbonbrief website, 11 May 2021; IEA Renewables 2021 report available at www.iea.org/reports/renewables-2021.

increased world prices will generally shift demand away from gas, while creating bumper profits on the supply side and potentially even more exploration. By mid-2022, it is not clear whether, on balance, the Ukraine crisis will hasten the global transition to renewable energy or stimulate additional investment in fossil fuels that may prolong their use in some economies. The dramatically lower costs of renewable energy will eventually drive out fossil fuels, but the risk is that the Ukraine crisis may slow the rate of decline at a critical time for global climate protection.

Similarly, it is unclear what the heightened emphasis on national security in energy supplies will do to Australian opportunities in the global energy transition. Australia's best long-term hope to maintain

Figure 2.2 Historical and prospective fossil and industrial CO_2 emissions. The coloured bands indicate scenarios investigated by climate change research, with the most recent shared socioeconomic pathways indicated in dark grey.

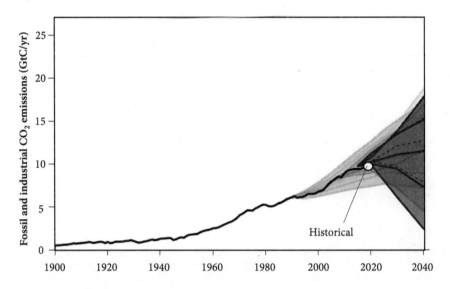

Source: Adapted from IPCC WG1 AR6, Figure 1.29.

its export-driven wealth is to export its renewable energy abundance in one form or another, including energy-intensive metals, fertilisers, high-voltage direct current (HVDC) lines into Asia or various hydrogen energy carriers on the world's seaborne market. However, desires in every world region for energy independence might reduce demand for low-emissions products from Australia. Conversely, Australia's relatively stable political environment (at least compared to other parts of the world) could increase demand for its energy exports.

Will the transition be swift enough to avoid catastrophic climate outcomes? Will net-zero carbon dioxide (CO_2) and then net-zero greenhouse gas emissions come soon enough? We have missed quite a few chances to improve the prospects for a happy ending. Some countries spent astronomical sums on COVID-19 recovery packages with little regard for the opportunitiy to accelerate transformations necessary to

Figure 2.3 Concentrations of the three main greenhouse gases: carbon dioxide, methane and nitrous oxide, over the last 2000 years.

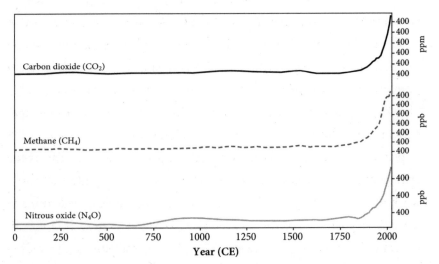

Source: Meinshausen et al., 2020 for year 0 to 2014 and World Data Centre for Greenhouse Gases for years 2015 to 2020.

Note: Greenhouse gas surface concentrations are compiled from Malte Meinshausen et al., 'Historical greenhouse gas concentrations for climate modelling (CMIP6)', 2017, available at: https://gmd.copernicus.org/articles/10/2057/2017/ for the years 0 to 2014; for 2015 to 2020, World Data Centre for Greenhouse gases, available at https://gaw.kishou.go.jp/publications/global_mean_mole_fractions.

minimise disruption from a changing climate. In Australia, the 'gas-fired' recovery promoted by the Morrison government artificially extended the life of fossil fuels. After the downward blip of about 6 per cent in global CO_2 emissions in 2020 resulting from COVID-19 restrictions,[3] global emissions in 2021 set new world records[4] (see Figure 2.2). Methane concentrations jumped to their highest level in human history – 1890 parts per billion (ppb) – fuelled by leaky fossil-fuel infra-structure. CO_2 concentrations now stand at 416 ppm (Figure 2.3).

Compared to 2015, when the Paris Agreement was adopted, 2021 marked a further increase in ambition. The 2021 United Nations Climate Change Conference in Glasgow (COP26) adopted text that strengthened its focus on limiting warming to 1.5°C. Targets and policies aimed at 2°C are no longer regarded as sufficient. In this chapter, we look at national emission reduction pledges and whether they are consistent with that 1.5°C ambition. It will come as little surprise to learn that they are not. But national commitments to reducing emissions, if met in full and on time, are, for the first time, projected to hold temperature rise below 2°C. The chapter then outlines what is required if pledges are to be consistent with limiting temperature rise to 1.5°C. All the options for keeping 1.5°C in play necessitate faster emissions reductions before 2030. This is the critical decade. Beyond the global picture, this chapter offers a framework which can be used as a guide to when countries need to reach net-zero emissions in order to keep the goal of 1.5°C alive. In line with generally accepted principles of fairness, developed countries, including Australia, will need to achieve net zero in the 2040s. The more prosperous of the other economies, including the group that committed themselves to net-zero emissions by 2060 at Glasgow (Russia, China and Indonesia – and many countries in Southeast Asia) will need to achieve net zero in the 2050s. India and other developing countries would need to achieve net zero in the 2060s – a decade earlier than India's commitment of 2070 at Glasgow.

2022 should be a wake-up call for all countries to accelerate emissions reductions. The call is loudest for Australia, the laggard among developed countries. In this chapter, we outline why the Morrison government's targets fell far short of what is required to limit warming to 1.5°C and to meet its commitments under the Paris Agreement. The Albanese government's adoption of a 43 per cent target is a first step in the right direction, but is nonetheless inconsistent with a journey towards 1.5°C. While the economic benefits that can flow from low-cost renewable energy will carry us a long way towards lower emissions, strong policies are needed to get Australia most of the way to net zero by 2035. To reach net zero in the 2040s, action this decade remains critical.

2030 climate targets

One hundred and ninety-six countries officially submitted climate pledges for 2030 in the lead-up to COP26 in Glasgow. These are called 'nationally determined contributions' (NDCs). The name emphasises the fact that each country autonomously determines its fair contribution to achieving the common targets of the Paris Agreement. There is no single formula which determines each country's fair share of emissions reductions.

The joint goal of all parties is enshrined in Article 2 of the Paris Agreement: Holding global average temperature rise to well below 2°C above pre-industrial levels and pursuing efforts to limit the temperature increase to 1.5°C above pre-industrial levels. In the past, the goal has been interpreted by some as 'aim to stay below 2°C, although some countries prefer to keep warming below 1.5°C'. The Glasgow Climate Pact strengthened the language around the 1.5°C target. It is now widely interpreted as a single goal of pursuing best efforts to limit warming to 1.5°C and if there is a temporary overshoot, ensuring that we nonetheless keep well below 2°C at all times.

Do all the individual, nationally determined 'fair' shares add up to an emissions trajectory in line with the global goal?

The old view of negotiations was akin to a piñata at a kids' birthday party. Before the game begins, the kids agree to take only their fair share of lollies. A papier-mâché dinosaur hangs from a tree, filled with lollies. Energetic, blindfolded five-year-olds beat it with a stick until it bursts open to reveal its sweet treasures. When it does, the frenzy reflects the uncontrolled self-interest of the parties. The 'fair share' for each one of them now is their own definition of what is 'fair' – probably more akin to 'I feel unfairly treated if I don't get the most lollies'. The tragedy of the commons.

Luckily, this framing of the climate policy dilemma is no longer apt. One change derives from the self-interest of governments to be re-elected – or, in autocratic states, to retain legitimacy. Electorates' desire for climate action has strengthened, driving towards greater ambition in nationally determined contributions. Second, in the international arena, the piñata game is played repeatedly, and the greedy in early games lose out in later rounds. A winning strategy in the repeated game is to 'be nice, be clear, be retaliatory, be forgiving'[5]. Third, the cost advantages of renewable energies and the opportunities in the energy transition now mean that moving earlier and faster can reduce energy costs, grow new industries and jobs, improve domestic energy security and help avoid the cost, environmental impacts and social turmoil of unabated climate change.

The NDCs at Glasgow fell far short of what is required in the crucial years remaining until 2030. Global emissions will stay roughly constant until 2030 if we add up all the national climate pledges. This is far from the 45 per cent reduction in CO_2 emission reductions relative to 2010 that the IPCC special report on global warming of 1.5°C found to be in line with a minimum-cost path to 1.5°C. Some countries' NDCs incorporate a rise in per-capita emissions to 2030: Turkey, a relatively high-capacity emerging economy, proposes a per-capita increase of 83 per cent from 2015 to 2030 – beyond a reasonable emissions projection even in the absence of climate policies.

India, which has very low per-capita emissions (around 2.4 tonnes of CO_2 equivalent per person (tCO_2eq/cap) in 2015), has targets that imply an increase of a third by 2030. Its solar energy capacity target of 100 GW, set for 2022, was ambitious and India is going to achieve only three-quarters of it due to too few rooftop solar installations.[6] Installation rates will have to pick up significantly if the new target of 450 GW renewable energy capacity installation by 2030 is to be achieved. Going faster is likely to reduce energy costs in India and many developing countries: new coal power in India faces an uncertain future given the falling costs and decentralised advantage of solar PV.

High-capacity, high-income countries generally submitted updated NDCs in the lead-up to COP26 that strengthened the pace of per-capita and total emission reductions targets for 2030. Coming from very high per-capita emission levels, the United States put forward a 50 to 52 per cent reduction in total emissions below 2005 levels by 2030. The EU, now surpassed by China in its per-capita emission levels, is aiming for 55 per cent below its 1990 levels by 2030. Australia remained an outlier, despite even higher per-capita emission levels than the United States. The Morrison government's target for 2030 implied reductions about half those of the US; the picture is somewhat enhanced now under the Albanese government.

For a deeper understanding of what the NDCs imply for projected emission levels in 2030, we need to distinguish between the unconditional and conditional elements of the NDCs. Several developing countries put forward a second, more ambitious, tier of emission reduction targets that are conditional on external financial, technical and/or capacity building support. When we assume that all the conditions for support are met and all NDCs as they were submitted by 31 December 2021 are met in full and on time, 2030 emission levels range between 50.7 and 53.3 Gigatonnes of CO_2 equivalent emissions[7] ($GtCO_2$ eq).[8] That range is between 3.4 per cent lower and 1.7 per cent higher than the 2019 pre-COVID global emission levels of 52.5 $GtCO_2$ eq. If we remove all the elements of NDCs conditional on external support, meeting the 2030 NDCs would result in global emissions between 54.2 and 56.7 $GtCO_2$ eq, which is 3.2 to 8.1 percent higher than in 2019.

Some countries may overachieve their NDC targets. This may be due to costs being lower than anticipated. In some countries, security objectives may lead to local renewable energy replacing imported fossil energy. For others, including Turkey and Pakistan, emissions are unlikely to rise as much as their pledges imply even if they implement no additional climate policies. Their 2030 emissions pledges are substantially above 'business-as-usual' emission projections.[9] For countries in this position,[10] 2030 emissions are projected to be 2.4 to 3.8 $GtCO_2$ eq lower,[11] if they simply follow their business-as-usual emissions rather than pursuing the higher emission levels in line with their NDCs.

How do these 2030 emissions targets compare to economically efficient scenarios that would have a good chance of keeping temperatures to around 1.5°C and well below 2°C? Many scenarios are built around keeping costs of climate mitigation low. These scenarios feature strong mitigation between 2020 and 2030. That's a long way from the sum of the Glasgow targets. The IPCC special report on 1.5°C global warming highlights the necessity of reducing CO_2 emissions by 45 per cent by 2030 to reach 1.5°C with no or limited overshoot and relatively low costs. This was recognised in the Glasgow Climate Pact – the first time such a specific milestone found its way into a global agreement.

In fact, the stable emissions until 2030 that follow from the sum of the Glasgow 2030 commitments are consistent with the International Energy Agency (IEA)'s 'no-additional-climate policy' scenario. Driven largely by the fall in the costs of renewable energy IEA's 'business as usual' scenario now shows stable emissions to 2030.[12] This scenario is much better than what we considered 'business as usual' a decade ago, but still results in catastrophic warming of around 2.7°C (median projection) by the end of the century.[13] The IEA's stated policy scenario (STEPS), which represents the IEA's best estimate of the emission trajectory in line with policies currently being implemented, yields warming of around 2.6°C by the end of the century.[14]

The scenario where emissions are constant until 2030 is dramatically better than what was predicted in 2015, prior to the Paris Agreement. The worst-case-scenario futures are avoided. Under such a scenario, we would not continue the fossil-fuel addiction that would lead to warming this century of above 3°C or 4°C, which humanity was

Figure 2.4 2030 NDC targets compared to economically efficient scenarios that achieve well below 2°C, 1.5°C with limited overshoot or lower than 1.5°C warming based on scenarios assessed in the IPCC Special Report on 1.5°C warming. The progression shows the aggregate effect of the change in countries' climate targets between 2015 and the end of 2021.

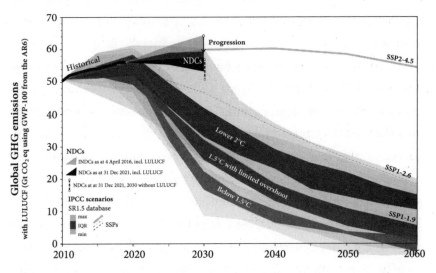

Source: Figure 5 from UNFCCC GST Synthesis Report UNFCCC Secretariat GST Synthesis report, March, 2022, available at: https://unfccc.int/documents/461517.

facing previously. But the 2030 emission targets and NDCs are woefully inadequate when measured against our common goals of limiting warming this century to 1.5°C and staying well below 2°C. The range of 2030 outcomes from full implementation of all conditional or unconditional elements of NDCs is hanging far above the cost-efficient scenario pathways we ought to be on (Figure 2.4). Those cost-efficient transition pathways are consistent with limiting warming to 1.5°C or 2°C.

The 2050 light on the horizon

In the lead-up to COP26, seventy-four countries, including all developed countries, put forward targets for mid-century. These long-term targets are often for net-zero CO_2 or net-zero greenhouse gas (GHG) emissions.

Article 4.1 of the Paris Agreement calls for net-zero GHG emissions in the second half of the century.

Those long-term targets are a light on the horizon. The per-capita emission levels in 2050 from countries that made long-term pledges are similar to those required under the 1.5°C or well-below-2°C scenarios assessed by the IPCC.

A symbolic milestone was achieved on 1 November 2021. The prime minister of India, Narendra Modi, went to the podium at Glasgow and announced a net-zero by 2070 target for India. This announcement meant that the world's targets, if implemented on time and in full, would limit projected warming to 2°C. India's announcement following China's net-zero CO_2 target for 2060, reduced our best estimate for end-of-century temperature rise to around 1.8°C[15] or 1.9°C, if all countries' commitments are met on time and in full.[16] While not *well* below 2°C in line with the Paris Agreement, it is nevertheless a clear sign of increasing ambition.

The long-term targets will not be achieved by themselves, and most are not underpinned by clear and transparent policies to carry us to net zero. In particular, the 2030 targets are insufficient and in most cases not in line with a trajectory towards the longer-term net-zero targets. There is some hope that the outlook for 2030 will improve thanks to the Paris Agreement's design, which creates pressure on all countries to regularly resubmit, revise and strengthen their targets to be in line with the highest possible ambition.

The role of offsets in a net-zero world: Correcting a deep misunderstanding

Net zero is referred to as 'carbon neutrality' by many parties today. Several voluntary climate accreditation schemes and business reports today simply subtract various 'offsets' from emissions to produce a big zero and then claim that they are 'carbon neutral' or 'climate neutral'. However, on our journey towards net-zero emissions, this deduction is only legitimate for offsets that genuinely and permanently remove carbon from the atmosphere.

Many 'offsets' today only reduce emissions relative to some assumed, counterfactual baseline level. Such schemes do not permanently reduce the stock of carbon in the atmosphere. For example, in some schemes, a gas producer may be able to generate offsets (receive carbon credits) by capturing and storing a proportion of its fugitive emissions. The whole activity of producing gas and sequestering some or all fugitive emissions does generate less emissions than it did before, but it still increases atmospheric carbon. These improvements in efficiency do not actually remove carbon from the atmosphere so cannot (with any scientific justification) be used to offset activities that release carbon into the atmosphere elsewhere.

The following example is taken from the current practice of the Australian Emissions Reduction Fund (ERF). An operator of an underground coalmine receives credits for burning the methane emissions that creep out of the underground mine. Once burned, 'only' CO_2 is emitted into the atmosphere, instead of the much more harmful methane. Thus, the coalmine operator receives credits. Those credits are given even though emissions still occur. Those credits can then be used to offset other corporate emissions. The mine operator receives Australian carbon credit units (ACCUs) and the purchaser of the credits, if it buys enough, can brand itself as carbon neutral under the government-sanctioned 'Climate Active' scheme. Alternatively, if the mine operator were to use those ACCUs they could assist it to meet what is required to claim to be carbon neutral, even though its operations still emit CO_2. Offsets of this type do not deliver the removals of CO_2 from the atmosphere required to meet the goal of achieving net-zero.

In a net-zero world, 'offsets' are only legitimate if CO_2 is removed from the atmosphere and stored permanently. This could be via CO_2 removal technologies such as mineralisation or direct air capture and storage. It could include bioenergy combined with carbon capture and storage (BECCS): carbon from sustainably grown biomass or biomass waste products being buried in geological formations (either after burning or directly). Or it could include char from pyrolysis of sustainably grown or waste biomass being used to add permanently to the soil's carbon content. Natural growth of plants through photosynthesis involves drawing on solar radiation and CO_2 from the atmosphere. When plants are harvested, if their carbon content is captured and sent to permanent storage in geological structures or soils it adds up to permanent reduction of the amount of carbon in the atmosphere. BECCS and the addition of char to soils (although there are caveats, particularly related to scale) could be important contributors to net zero. To avoid the risk of negative side effects, the biomass should be waste from the processing of agricultural products, or grown on land that does not compete with food production or grown on land that does not have high biodiversity value that would be lost.

CO_2 removal technologies must provide a permanent and additional store of carbon. Offsetting fossil fuels with legitimate biospheric carbon storage is potentially risky. The biospheric stores need to be around for hundreds or thousands of years to offset emissions from fossil fuels that were locked away in the geosphere until being extracted and burned. For forests, this requires protecting the store of carbon from harvest,[17] beetle attacks and disease, drought and wildfire, which may be increasingly difficult in a changing climate. There is also an issue with additionality associated with many biospheric stores: it is necessary to be confident that the overall quantity of forests is increased relative to what would have been there anyway over these long/permanent timeframes. Biospheric offsets have a limited role to play in achieving a net-zero emissions world, but they should also be treated with caution. Only putting carbon back into the geological reservoirs where the carbon came from (or mineralisation) comes with a high likelihood of permanence.

Figure 2.5 Greenhouse gas emissions and twenty-first century temperature implications of 2030 NDCs and long-term targets (as of 12 November 2021).

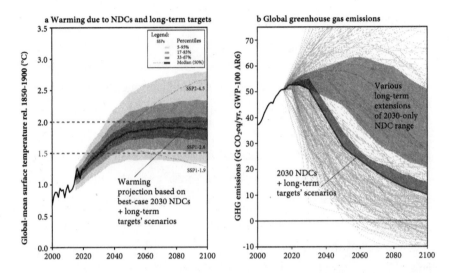

Note: Warming projections consistent with IPCC WG1 AR6 science are shown for the best-case emission trajectory in line with 2030 NDCs and long-term targets. Not relying on 'best-case' interpretation of targets, but assuming all conditional targets to be implemented would yield median warming just around 2C (not shown). Greenhouse gas emissions in line with the best-case (solid line at bottom of NDC plume) and not best-case (line at top of NDC plume) as well as various extensions on the basis of 2030 NDCs are shown in the context of the SR1.5 IPCC scenario database (panel b).

Source: Adapted from Figure 3 in Meinshausen et al., 'Realization of Paris Agreement pledges may limit warming just below 2 °C'.

We estimate that world emissions would fall by roughly 52 per cent below 2019 levels by 2050 for GHGs[18] and 58 per cent for CO_2, if NDCs and longer-term net-zero targets submitted to the UNFCCC were met on time and in full. There is still a gap between these targets and what would be necessary to achieve a global goal of net zero CO_2 by 2050. The second challenge is implementation: the long-term targets must be underpinned by policies that lead to the targets being met.

The methane pledge and other global initiatives

International pledges under the NDCs and long-term low-emission development strategies (LT-LEDs) are not the only targets in the climate arena. Many states, regions and cities have their own targets. In Australia, all states and territories had net-zero targets in various legal forms before Australia as a whole adopted the net-zero target for Glasgow.

Bilateral and plurilateral agreements often contain commitments to climate-relevant actions beyond members' NDCs and long-term pledges under the UNFCCC. The November 2021 COP26 summit in Glasgow resulted in a wider range of other commitments than previous international climate summits, including on forests, cars, coal phase-out, finance and methane. However, not all these commitments were new and careful analysis is required to determine how much additional climate benefit (on top of NDCs) they bring.[19]

The US–EU-led Global Methane Pledge was applauded at COP26.[20] Participants agree to make voluntary contributions towards a collective effort to reduce global methane emissions by at least 30 per cent from 2020 levels by 2030. Over 100 countries had joined by mid-2022, representing around 50 per cent of global anthropogenic methane emissions and over two-thirds of global GDP.

While methane reductions are important, the long-term impact of CO_2 in the atmosphere makes it crucial that action on methane does not detract from action to reduce CO_2 emissions as fast as possible. The Global Methane Pledge was announced as reducing 'warming by at least 0.2°C by 2050, providing a crucial foundation for global climate change mitigation efforts'.[21] When avoiding double counting in quantifying the emissions reductions, the overall additional benefit of the Methane pledge might be considerably smaller: some CH_4 reductions are already covered by the NDCs and long-term targets. Stripping out these impacts suggests the incremental impact of the Global Methane Pledge might be as little as ten times less: that is, 0.02°C by 2050. As a double-stitched security, the global methane pledge supports commitments under the NDCs and long-term targets. Its incremental contribution would be more important for countries with weak 2030 targets – but the developed country with the weakest target, Australia, has so far declined to join the Methane Pledge.

Large bilateral initiatives also emerged from COP26 in Glasgow, some of which have faced criticism that they may not live up to the hype that the media releases created. The Glasgow Climate Pact did not reference those initiatives.

However, 'clubs of the willing' can provide much needed impetus to strong action. The border tax adjustment mechanism agreed by the European Union in March 2022 is a good example of one such initiative.[22] Any imports of steel, cement, fertilisers, electricity and some other products into the EU will face an equivalent carbon price to these industries within the EU from 2026. Border tax adjustments by the EU require exporting countries such as Australia to choose between implementing a carbon price and having the EU earn tax from Australian products via the border tax adjustment mechanism.

What is the remaining global carbon budget?

Throughout and after 2009, a string of publications established that the 'sustainable level of emissions' for climate stability is basically zero.[23] Before that, the prevailing misunderstanding was that because nature soaks up some of the carbon we emit, warming could be stopped by reducing emissions to that level of natural uptake. However, temperatures will continue to increase until net emissions fall to basically zero. The old argument neglects the dynamic nature of the natural sinks and the Earth system's inertia.

Halting the CO_2-induced temperature increase requires not only stopping the increase in CO_2 concentrations, but actually reducing them. The Earth features a somewhat delayed warming in response to increased greenhouse gas concentrations. With constant CO_2 concentrations, temperatures would rise for centuries. Oddly enough – and it is just a coincidence – the rate of decrease in CO_2 concentrations that we require to stop CO_2-induced temperatures from rising matches the rate of decrease in CO_2 concentrations that occurs if CO_2 emissions are net-zero.

Derivation of the remaining carbon budget

The IPCC special report on 1.5°C and Working Group I's contribution to the Sixth Assessment Report employed a useful, and relatively simple, approach to deriving a specific carbon budget for a certain warming limit:

Remaining carbon budget = $(T_{target} - T_{non-CO_2} - T_{ZEC} - T_{hist})/$ TCRE $- E_{other}$

The 'remaining carbon budget' is the amount of cumulative carbon emissions from today that is consistent with limiting warming to T_{target}.

Apart from CO_2, humans also influence the climate via other greenhouse gases and aerosols. Non-CO_2 greenhouse gases like methane and nitrous oxide warm the climate too while aerosols are cooling agents that have cooled the climate over the last century. Reductions in emissions of non-CO_2 greenhouse gases cool the climate while reducing aerosol emissions leads to warming (via an unmasking of previously masked greenhouse-gas-induced warming). The balance of warming and cooling influences from non-CO_2 emissions are summarised in the T_{non-CO_2} term.

T_{ZEC} refers to the amount of CO_2-induced warming that we expect to experience even after having reduced CO_2 emissions to net zero.[24] This term captures many of the second-order effects that aren't captured by the first-order linear relationship between warming and cumulative CO_2 emissions. T_{ZEC} is highly uncertain and scientists are unsure whether it will be slightly positive or negative and to what extent it will also depend on the overall level of warming.

The historical temperature, T_{hist}, is subtracted from our target temperature, T_{target}, as we are interested in the remaining carbon budget from today. The significant warming we have already witnessed means there is less room for future warming.

The last term, E_{other}, is a correction factor, accounting for otherwise unrepresented carbon cycle or other feedbacks. We simply subtract this correction term to obtain our remaining carbon budget.

Overall, the remaining carbon budget is the amount of CO_2-induced warming that we can still afford, converted into cumulative CO_2 emissions via the TCRE. The TCRE (full name 'transient climate response to cumulative emissions of carbon dioxide') quantifies the Earth system's sensitivity to cumulative CO_2 emissions: that is, how much warming we get for each tonne of CO_2 we release. In IPCC AR6 WG1, the TCRE was assessed as having a likely range between 1.0°C and 2.3°C warming per petagram of carbon emissions (°C/PgC). The TCRE's uncertainty distribution is usually assumed to be symmetric, leading to a central value of 1.65 °C/PgC (this assumption was used in IPCC WG1 AR6 to calculate a remaining carbon budget for a 50 per cent chance of staying below 1.5°C).

The IPCC Sixth Assessment Working Group I (IPCC AR6 WGI) report states that for a 50 per cent chance of staying below 1.5°C of warming, the world has another 500 $GtCO_2$ to spend from the beginning of 2020.[25] That's just a bit over twelve years at current emission rates of 40 $GtCO_2$ per year. To be true to the Paris Agreement's temperature goal, which is relative to pre-industrial temperatures, we ought to start from 350 $GtCO_2$ as our remaining carbon budget from 2020 onwards (see Appendix 2.1). That means not twelve years but just nine years of carbon emissions at current levels.[26] The carbon budget would be even smaller if the role of non-CO_2 forcers is larger than assessed in the IPCC AR6 WGI report.[27]

Whether nine or twelve years, the bottom line is that if we do not start to drastically reduce global CO_2 emissions in the next few years, we will have blown the 1.5°C carbon budget by the end of this decade. In fact, all scenarios investigated by the IPCC project reach around 1.5°C warming by the early 2030s. The question is whether or not our activities this decade set the world on a course to shoot dramatically past that warming.

To halt temperature rise, we must stop pumping more and more carbon from the inert fossil carbon reservoirs into the active carbon pools. The carbon cycle can be thought of as three big bathtubs connected via hoses. The more water (carbon) you put in one, the higher the level in the other tubs over time. The active carbon pools that are the three interconnected bathtubs are the atmosphere, ocean and terrestrial biosphere. The inactive and inert reservoir, which is normally not directly linked to the other three, is the geosphere, with all its fossil carbon. A little bit of mineralisation and rock weathering occurs, but until humans poked a large hole into that secure store, the carbon contained in the geosphere hardly interacted with the active pools.

The remaining carbon budget is the amount of CO_2 emissions that we can still pump into the atmosphere without exceeding a given temperature goal, with some limited adjustment for those CO_2 removal technologies that can lock additional carbon away for hundreds of years (see 'The role of offsets' above). If we overspend on our budget, we commit future generations to drawing that CO_2 back out of the atmosphere. Such an overspend also comes at the price of overshooting the temperature target.

As of 31 December 2021, the NDCs amounted to approximately 437 $GtCO_2$ of emissions between 2020 and 2030. A carbon budget consistent with 1.5°C – 500 $GtCO_2$ – leaves only 63 $GtCO_2$ beyond 2030[28] – roughly one and a half years of current emissions.[29]

Figure 2.6 Global greenhouse gas emissions to meet the Paris Agreement goal of net-zero GHG emissions in the second half of the century and a pathway close to 1.5°C warming.

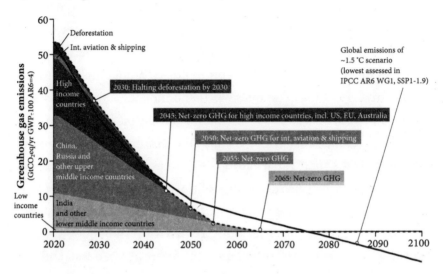

Notes: Compared to the illustrative regional emission wedges, the lowest scenario investigated by the IPCC Sixth Assessment Working Group I, the so-called SSP1–1.9 scenario, has slower reductions after 2050 and net-negative GHG emissions after 2075. The global and aggregate illustrative phase-out wedges closely match the SSP1–1.9 trajectory. They consist of halting deforestation by 2030, aiming for net-zero GHG by 2045 among 'high income' countries, moving along a straight line trajectory for a phase-out by 2055 for so-called 'upper middle income countries' (like China and Russia), and to net-zero GHG by 2065 for 'lower middle income' and 'low income' countries. Note that these wedges are not meant to indicate 'fair' shares, but rather illustrative territorial emissions. A fair distribution of mitigation action is both related to territorial emissions as well as any transfer flows of credits, finance or other implementation support.

Source: Authors' calculations based on PRIMAP-histCR for historical emissions, UN Population prospects for current population, RCMIP for SSP1–1.9 GHG emissions, World Bank Income grouping categorisation.

A range of factors renders the remaining carbon budget uncertain. Non-CO$_2$ emissions could be higher or lower, depending on what policies and actions are taken to reduce emissions. The effect of non-CO$_2$ warming greenhouse gases and cooling aerosols could also be different to the default assumption made in the IPCC Sixth Assessment (Working

Group 1) report. While most of the known Earth system feedbacks are included, carbon cycle, permafrost and clathrate feedbacks could be higher or lower than projected.

Aside from the remaining carbon budget, the IPCC Sixth Assessment Working Group III report provided emission milestones for its scenarios that meet 1.5°C warming with no or limited overshoot (Figure 2.6). The key milestones are emissions reductions of 37 per cent below 2010 levels by 2030 and net-zero CO_2 by 2050. The lowest scenario investigated by the IPCC Sixth Assessment Working Group I report, the so-called SSP1–1.9 scenario, is very close to these emission milestones. It turns out that the SSP1–1.9 scenario has around 565 $GtCO_2$ emissions between 2020 and 2050, the point when total CO_2 emissions reach net zero in that scenario. The slightly higher cumulative emissions of 565 $GtCO_2$ is roughly in line with the IPCC Sixth Assessment Working Group I report finding of a 500 $GtCO_2$ budget, given that the SSP1–1.9 scenario also includes a limited overshoot of the 1.5°C level, before going lower than 1.5°C by the end of the century.

There are many ways to slice and dice a global greenhouse gas emission pathway that would limit warming to around 1.5°C. The partitioning into 'developed' and 'developing' country emissions is superseded by world realities in many respects. At the Glasgow summit, China and Indonesia distinguished themselves from other developing countries with commitments to net zero by 2060. Russia adopted a domestic long-term strategy (although it did not not yet officially submit this to the UNFCCC), which includes a net-zero GHG target for 2060. Given the generous (and creative) land-use accounting by Russia, the target is in effect rather weak, and trust in Russia's stance towards international multilateral frameworks is unquestionably shattered after the invasion of the Ukraine. Nonetheless, using the World Bank's income group categorisation allows us to form three groups: the high-income countries, including the United States and the European Union;

the upper-middle-income countries, including Russia and China; the lower-middle-income countries, such as India; and finally, and also in terms of emissions much less important grouping, the 'low-income' countries. If the five following elements can be achieved, the world has a chance to keep warming to around 1.5°C (with no or limited over-shoot), with slightly falling temperatures in later years:

1. **Halting net deforestation globally by 2030.** At the COP26 Glasgow climate summit, leaders of more than 100 countries, including Brazil, Indonesia and China, repeatedly pledged this goal.[30]

2. **Reducing international aviation and shipping emissions to net zero by 2050.**

3. **Reducing 'high income' country emissions to net-zero greenhouse gas emissions by 2045.** That is five years earlier than most phase-out targets at the moment, such as those from the United States and European Union.

4. **Bringing forward the 'upper middle income countries' net zero GHG commitments to dates in the 2050s.** Note that China's net-zero target is currently for CO_2, not GHG. A Chinese target of net-zero GHG by 2055 would be both a strengthening to include all GHGs and bringing forward the net-zero target by just five years.[31]

5. **Reducing the 'lower middle income' and 'low income' country emissions to net-zero GHG emissions in the 2060s.** Note that a longer phase-out by very low per-capita emitters such as India could be compensated by earlier phase-outs for the high-income countries. Also note that the plotted wedges are illustrative. In particular, for these lower two groups, a plateauing or initial limited further increase of emissions over the next decade might better reflect fairness principles and real world dynamics.

Figure 2.7 Expected global temperatures (medians and 5 per cent to 95 per cent ranges), if all NDCs and long-term targets as of 31 December 2021 are realised (scenario 'A' in Meinshausen et al., 2022) as well as the lowest scenario investigated by IPCC Sixth Assessment Working Group I, namely SSP1–1.9, which is in line with the '1.5°C with no or limited overshoot'. Shown temperatures are derived using our IPCC Sixth Assessment Working Group I calibrated climate emulator MAGICC7.5 as described in Cross-Chapter Box 7.1 in IPCC AR6 WG1, available at www.ipcc.ch/report/ar6/wg1/.

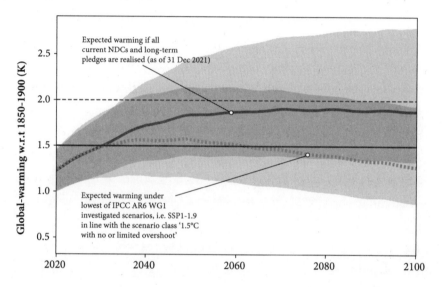

In the example presented in Figure 2.6, 24 per cent of emissions originate in the high income counties from 2020 onwards, representing approximately 14 per cent of the world's population. Thus, the illustrative allocation of emissions targets in Figure 2.6 would give high-income countries more than they would get under a rule of 'equal per-capita emissions from now on'. Of course, that share is also biased towards 'high-income' countries, if you were to consider 'equal cumulative per-capita emissions' since some historical date as a fair distribution of emissions. Given the much lower costs of zero emission technologies, there is luckily not the same 'mitigation burden' as there was a few years ago. Still, weaning off existing fossil fuel infrastructure at high speed can be painful so support and incentives are required. The discussion of a 'fair' distribution of emissions reductions will be ongoing. The almost inevitable mismatch between physical territorial emissions and the 'fair share' of emissions could be overcome by traded carbon credits or higher financial contributions to deal with climate impacts. For a net-zero future, carbon credits for negative emissions are the only way to truly 'offset' higher emissions (see breakout box on page 117).

This section has provided one possible way to consider the mitigation challenge we face and how to differentiate countries on the basis of development and capability. However, there are other ways to think of this challenge. One is to explicitly divide up the remaining carbon budget.

How to slice up the global budget

Calculating remaining carbon budgets for individual nations involves value judgements regarding who deserves how much. It's again like kids at a birthday party. This time the best analogy is the birthday cake. Let's imagine that early on in the party, one group of kids devoured half the cake. Later, additional sweets were brought to the table (the renewables that can fulfil the kids' energy needs), but the question is how much of the 'fossil' cake should be given to whom?

The most straightforward solution is to make sure every kid at the party gets an equal amount of cake overall. Those who have already eaten a lot of the cake get less, if anything, later on. In climate negotiations, this is the proposal by India and others, who support an 'equal cumulative per capita approach'. The birthday cake is not quite the same as in the climate negotiations: today's past and future emitters are not the same people, but the grandparents and descendents of today's emitters. The argument is however that the fossil-fuel consumption of our ancestors provided a number of economic and development benefits to those who came later. In simple terms, the main argument is that every person should get the same amount of cake, and if ancestors ate more than their fair share, the kids should make up for it and eat correspondingly less.

There are some kids, such as the developed countries, who don't like this way of dividing the remaining cake but know that their argument is morally weak. Over the last 100 years, they (or their ancestors) enjoyed economic development with high per-capita emissions and have a fossil-fuel infrastructure that still churns through much fossil carbon these days. They argue that they should be able to keep eating more per person now, and for everyone to eat the same amount per person at some point in the future. In climate parlance, that is often called 'per-capita convergence', or 'contraction and convergence'.

You can imagine the screams of the kids that did not have the same amount of cake in the past and are being told that the rich kids are for the time being still allowed to eat more per person than them. Why should they reduce their intake from now on to end up at the equitable state of equally sharing the remaining crumbs at the end of the party?

The Garnaut Climate Change Review (2008), followed by the Australian Climate Change Authority (2014), sought to bridge the fairness gap by accepting that the convergence towards equal per-capita emissions may start later for developing than developed countries. Developing countries would not be required to converge downwards until their per-capita emissions had reached the declining average level of the developed countries. Kids who had eaten little birthday cake until now could increase their consumption in the near term, while those who had been consuming the most start reducing right away. While such a 'modified contraction and convergence' might seem fair and the only realistic approach from a rich-kid perspective in Australia, it is little surprise that some less developed countries do not find it attractive.

There are further considerations when one attempts to define a fair distribution of the remaining emissions space.

First, all fossil-fuel emissions cause harm to everyone living on this planet. An 'equal cumulative emission per-capita approach' is often suggested as a fair approach. Allowing everyone to do the same kind of harm does not necessarily get us to the best possible place in the future, though. That is not to say that there is no responsibility for harm done in the past and present.

Second, it is energy services, not emissions, that fulfil our human needs. We can satisfy our heating needs at unnecessarily high cost by using coal, like our grandparents, in badly insulated houses. Alternatively, we can do so by using highly efficient renewable-powered electric heat pumps in insulated houses. Returning to the birthday party metaphor, new technologies start to question the argument for 'the same amount of cake for everyone'. Maybe, if there is something else but chocolate cake emerging on the table, a deal can be made that everybody gets cake, although not necessarily chocolate cake? While it's true that the early fossil-fuel cake was largely eaten by one group of kids, as there are now many sweets and cakes to go around, overall fairness lies in enabling access now to all kids to enjoy these while trying to cut down on eating any more of the original 'fossil' cake.

The other fundamental issue is that there is no sustainable level of emissions and there is no greenhouse gas emission that does no harm. As a result, a corrective justice approach is essential. This can become extremely delicate in negotiations. Countries with high historical emissions have high responsibility for the climate damages that are and will continue to be unleashed on us. It is generally these same high-per-capita emitters that have the capacity to support the global energy transition and adaptation actions, and to financially support those who could or did not adapt to climate change and suffered 'loss and damage'.

Reaching agreement on the appropriate way to approach these issues is never easy. In the 2000s, there was a big push in the climate negotiations for the so-called 'Brazilian proposal'. Considerable scientific effort was undertaken to devise methods to calculate the temperature contribution from individual countries[32]. Equity approaches around the historical 'polluter pays' principle seem straightforward yet rarely are in practice and our kids' birthday party analogy illustrates one reason why. Under the Brazilian proposal, those who ate most of the original cake are also responsible for cleaning up the sticky mess. That of course can make those early fossil carbon kids (those who still have cake crums smeared around their mouths) scream. The US screamed, insisting in the Paris Agreement negotiations that a clause be introduced that renders the 'loss and damage' language (seen by some as a major advance in international negotiations) completely ineffective. The US clause says that the loss-and-damage text is no precedent and should not form the basis of any compensation.[33]

The kids' cake analogy also hints at how interconnected and complex the question 'what is fair?' is in climate negotiations. Mitigation is linked with adaptation, finance, transparency, and loss and damage discussions – all seen from different equity perspectives by different stakeholders. A right to more emissions now seems a perverse starting point for most of the equity approaches. As noted above, new coal power plants now make little economic sense and cause environmental and public health damage at the same time. A corrective justice approach that takes into account both the harm inflicted by past emissions and the opportunities from low-cost renewable energy provides a complementary and possibly better lens. Nonetheless, approaches to quantifying shares of the remaining carbon budget can still provide a useful guide andare discussed in the next section.

Formulas for the way forward

The quest for a formula that provides a quantitative and equitable solution to sharing the burden of reducing emissions is as old as the climate negotiations themselves – at least for observers and some of us academics. Governments in negotiations have not actually been seeking that one formula. Academics and some NGOs put forward proposals, but quantified approaches only provide the backdrop against which the geo-political horse-trading takes place. Raúl Estrada-Oyuela, the Argentinian ambassador chairing the Kyoto Protocol negotiations, did not go around from delegation to delegation asking which formulaic equity approach they would prefer. He went from delegation to delegation asking: What do you propose for your emission reduction number in Annex B? Governments had to balance the scrutiny by their people, commitment to being good global citizens in the fight against climate change, and the domestic and international blame that will ensue when proposing targets that do not even pretend to reduce emissions (such as Australia, with its 108 per cent targets in the first round of the Kyoto Protocol in 1997).

Nevertheless, formulaic approaches between different countries do guide mitigation action in some contexts, such as within the European Union. What is considered 'fair' has many dimensions, so internal EU approaches are complex. Political will and feasibility, capacity and other national circumstances are conflating factors. That has led the EU to devise highly complex approaches to equity, such as the so-called 'tryptich approach'.[34]

In other contexts, quantified equity approaches are useful guardrails for assessing whether targets are aligned with the Paris Agreement.[35] Below, we outline the range of equity approaches used, with 'equal cumulative per capita' at one end and 'linear convergence to equal per-capita emissions' at the other. India and other low-emitting developing countries argue for cumulative per-capita emissions; developed countries tend to favour a transition towards equal per-capita emissions. The latter introduces a form of grandfathering: those with high current per-capita emissions start with higher emission allocations.

Towards the middle of this range is equal global average per-capita emissions from today onwards. In 2021, the German Advisory Council on the Environment (SRU) considered how a remaining global carbon budget could be allocated among nations. They proposed that Germany could use as much of the remaining carbon budget as its world popula-tion share – that is, 1.1 per cent. They started with an assumption that a 1.75°C temperature level with 50 per cent likelihood can be considered equivalent to a 'well-below 2°C'. Using the corresponding remaining carbon budget for 1.75°C, the SRU then translated this into a 2045 net-zero year for EU-28 CO_2 emissions, five years earlier than previously envisaged. In a landmark case, the German High Court[36] demanded new targets after 2030, which led to a net-zero GHG emissions year being turned into law for Germany. This was swiftly implemented by the outgoing coalition government under Angela Merkel.

Australia's world population share is 0.33 per cent. If the approach applied in Germany was used, Australia's share of the remaining car-bon budget would therefore be 0.33 per cent. The Australian carbon budget implicit in the commitments taken to the Glasgow conference would exceed this allocation by a factor of more than four.[37] The mod-ified contraction and convergence proposed by the Garnaut Climate Change Review in 2008 and used by the Climate Change Authority (2014) adopted an estimate of Australia's fair share of 0.97 per cent of the global carbon bubget, exceeding the forward-looking population share by a factor of almost three.[38] Even acceptance of net zero in 2045 from –43 per cent in 2030 (the Australian Labor government's near-term target) would substantially exceed the Australian share for a 1.5°C or 1.6°C target.

Many of the Australian studies after Garnaut (2008) and the CCA (2014), including those to which we have contributed, adopted their approach – partly because the political precedent had been set, partly because the quantified range of other proposals (which also included some sort of grandfathering) was strewn around 1 per cent. In particu-lar, the Climate Targets Panel[39] and the Victorian targets panel (both of which we contributed to) adopt the 0.97 per cent share. If Aus-tralia insists on 0.97 per cent of the remaining cake, then somebody else has to swap to the renewable treats earlier or stay hungry (on a per capita basis). And yet officially Australia has never accepted targets as demanding as those proposed using the modified contraction and convergence approach to determining its fair share. Australia's strong economic as well as climate interest in the success of the global mitigation efforts warrants recondsideration of its approach.

In 2013, the IPCC categorised the myriad ethical principles that have been proposed in the scientific literature in its Fifth Assessment Report. We provided advice based on that to the Victorian government in 2018, quantifying various equity approaches to Australia's fair share of emissions until 2050.[40] As all these quantifications embodied a smooth trajectory from current emissions, they included a strong element of grandfathering.[41]

The difference between targets and actual emissions

At the time of the Kyoto Protocol, an unfortunate dynamic gripped international negotiations. Countries that expected or wanted large support from the international community for any mitigation effort (notably Russia and some Eastern European countries) set targets at unrealistically high emission levels. They hoped that this provided them with the opportunity to sell 'overachievement' of their targets. A similar approach seems to have motivated Turkey and Pakistan in Glasgow, as their targets imply an incredibly high and unlikely 'business as usual' emission growth.

Fortunately, the dynamic has evolved since the Paris Agreement. Most developing countries now put forward two targets, unconditional and conditional. The more ambitious the unconditional targets are, the more trust those providing the support for the conditional targets have that the developing country is seriously pursuing reductions in emissions. This conditional target indicates what stronger mitigation action would be possible if additional support were provided. The hope is that this twofold target encourages countries both to put forward meaningful unconditional domestic ambitions, and detailed domestic action plans that provide options for the deeper reductions that would be made possible by international support.

Where do country NDCs and long-term targets sit within this range of effort sharing approaches?

The NDCs and long-term targets currently submitted are not a reflection of any 'fair' approach to allocation. They are a mixture of realpolitik, different views of what equitable contributions are, different sets of political pressures and different understandings of the costs of making the transition.

Most of the developed or high-income countries currently propose 2050 net-zero GHG targets. The commitments made in the lead-up to COP26 and in Glasgow broadly follow, with an approximately five-year delay, the framework outlined in Figure 2.6 of higher-income countries pledging net-zero by 2050. Likewise, the announced and pledged long term targets of the more advanced other countries (most importantly Russsia and China) reaching that goal by 2060, with some countries with very low per-capita emissions, notably India, pledging to reach net-zero emissions by 2070. Thus, the three key imperatives for the international deliberations are: first to strengthen long-term strategies so that high-income countries in aggregate aim for 2045, China and other upper-middle-income countries for 2055, and India and the lower-income countries for 2065; second, to strengthen targets to consistently reflect net-zero GHG, rather than only net-zero CO_2 targets; and third, to advance near-term policies and actions so that the world is on a trajectory to these net-zero GHG targets. Many short-term crises provided (largely missed) opportunities (COVID, energy crisis due to Russia's war on Ukraine) and will offer additional opportunities to synergistically address multiple policy objectives, including mitigation action. On the other hand, if short-term crises over the next decade are turned into road blocks, we will delay the turn towards net-zero and push the goals of the Paris Agreement out of reach.

The gap between pledges and policies also needs to be addressed. In some countries, legislated policy packages were put on the table at almost the same time as the targets (see the European Climate Law in the EU). In other countries, like Australia, the net-zero target is not underpinned by any substantive work on how it will be achieved.

As discussed above, for the first time the sum of all pledges, if realised, would put the world on track to limit warming to just below 2°C with a 50 per cent chance. Given the climate impacts we witness at the current 1.2°C of warming, 2°C seems an undesirable destination. In the light of where the world was five or ten years ago, with pledges adding

Figure 2.8 The sum of climate pledges on the table by COP26 (2021). With the announcements of India and China, the world's projected median temperature dropped to just below 2°C on 1 November 2021, when India made its announcement about a net-zero 2070 target. The dark grey ranges indicate the projections based on conditional NDCs and the light grey range only includes unconditional NDC targets.

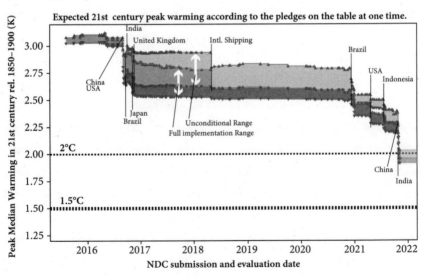

Source: Adapted from Figure 5b in Malte Meinshausen et al., 'Realization of Paris Agreement pledges may limit warming just below 2 °C', *Nature*, 604, 2022, pp. 304–9.

up to over 3°C warming, we have come a long way. However, there is still a long way to go to stay close to 1.5°C warming level and the crucial reductions must be made this decade. The 1.5°C carbon budget will be exhausted by around 2030 unless such immediate action is realised.

Australia's role

What does this mean for Australia's emissions reductions targets, if it is to contribute to putting the world on a 1.5°C pathway? Two issues need to be considered. First, we need to consider how different pathways for Australia's emissions compare to its fair share of a global 1.5°C emissions budget. Second, we should consider what role Australia can play in supporting other countries to reduce emissions by becoming a major exporter of low-emissions alternatives to fossil fuels and energy intensive goods.

The approach used to calculate the global emissions budget from the carbon budgets discussed in the previous sections are outlined in Appendix 2.1. This underpins the results presented in Table 2.1. In this section, we focus on the Australian share.

Australia's emissions budget

Multiple quantifications of Australia's fair share of the global emission budget can be found in the literature. Table 2.1 shows Australia's share and its emissions budget based on a selection of different possible assumptions of Australia's fair share of the global emissions budget, including an equal per-capita share, an equal cumulative per-capita share and the modified contraction and convergence approaches discussed earlier. Australia's fair share varies by a factor of nearly four (from 0.33 per cent to 1.27 per cent of the global emissions budget) depending on the equity approach adopted. This set of fair shares should be taken as an indication of different options (although we make no direct comment on which is more or less fair).

Table 2.1 Allocating the remaining GHG emission budget using various illustrative fair shares for Australia.

Fair share methodology	Australia's share of the remaining emissions budget from 2013 (%)*	Australia's remaining emissions budget from 2013 (GtCO$_2$)		
Temperature level and likelihood of staying below		<1.5°C with 50%	<1.6°C with 50%	<2.0°C with 83%
Global remaining emissions budget from 2013 up to 2050 relative to pre-industrial after LULUCF adjustment and international shipping and aviation are removed (GtCO$_2$eq)		853 GtCO$_2$eq	1013 GtCO$_2$eq	1279 GtCO$_2$eq
CCA (2014), based on Garnaut (2008)	0.97%	8.3	9.8	12.4
Equal per-capita 2040 convergence (Robiou du Pont et al., 2017 as used in Meinshausen et al., 2019)	0.73%	6.2	7.4	9.3
Equal cumulative per-capita (Robiou du Pont et al., 2017 as used in Meinshausen et al., 2019) – with net-negative emissions after 2050.	0.68%	5.8	6.9	8.7
Capability (Robiou du Pont et al., 2017 as used in Meinshausen et al., 2019)	0.52%	4.4	5.3	6.6
Greenhouse Development Rights (Robiou du Pont et al., 2017 as used in Meinshausen et al., 2019)	1.19%	10.1	12.0	15.2
Constant emissions ratio (Robiou du Pont et al., 2017 as used in Meinshausen et al., 2019)	1.27%	10.8	12.9	16.2
Australia's share in future emissions should equate its current share of world's population	0.33%	2.8	3.5	4.2

Superpower pathway compared to Australia's emissions budget

The pathway implied by the federal government's current targets for Australia and the Superpower pathway discussed in this book are illustrated in Figure 2.9.

The Superpower pathway discussed in Chapter 1 would see Australia comfortably overachieveing on its official target of –43 per cent by 2030, reaching –75 per cent in 2035 and zero net emissions in line with the rest of the developed world by 2045. That would be a dramatic change from Australia's past position, but one that would still not impress fair-minded observers in other countries. It would lead to cumulative GHG emissions of 10.1 $GtCO_2$-eq between 2013 and net zero. By 2030, a straight-line achievement of this Superpower pathway implies more than halving (–54 per cent compared to 2005) Australia's emissions.

On that pathway, Australia would not meet even its fair share for an 83 per cent chance of staying below 2.0°C under a quantification of equity based on capability (0.52 per cent) or 'equal-per-capita-from-now' (0.33 per cent). The Superpower pathway would be just above Australia's fair share of a 50 per cent chance of staying below 1.6°C, if we calculate Australia's fair share in line with the Climate Change Authority (2014): that is, assuming that Australia's 'fair' share of future emissions is 0.97 per cent, which is three times its population share (0.33 per cent). The cumulative greenhouse gas emissions under the 'Superpower pathway' would, however, amount to 23 per cent more than the Climate Change Authority (2014) fair share of 8.3 $GtCO_2$-eq (from 2013 onwards) for a 50 per cent chance of staying below 1.5°C.

---

* Note: the Australian emissions budget is shown from 2013, not 2020 or later. This is to enable comparison across all the 'fair share' approaches calculated, because 2013 is the date for which the widely used modified contraction and convergence estimate of the Australian fair share was derived by the Climate Change Authority (2014), based on Garnaut (2008). To get to an emissions budget at a future time, Australian emissions from 2013 to that future date would need to be deducted from the budget.

Figure 2.9 Australia's GHG emissions historically and in the future – assuming the CCA fair share of 0.97 per cent or the Superpower pathway considered in this book. The shaded triangle in the background for future emissions are the Australian GHG emissions in line with a 1.5°C with no or limited overshoot (up to 1.6°C with 50 per cent chance) – with the dashed line indicating a 83 per cent chance to stay below 2°C (upper bound) – again, while assuming that Australia's fair share of global emissions is three times its current world population share of 0.33 per cent.

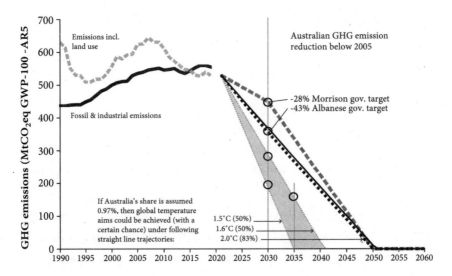

Source: Authors' calculations based on Australian GHG inventory data.

In comparison a –43 per cent target by 2030 and straight-line achievement of net-zero by 2050 would amount to cumulative emissions that are 46 per cent more than the 1.5°C (50 per cent) share.

Implications of Australia doing its part

The recent historical legacy makes it extremely challenging for Australia to stay within its fair share of a 1.5°C emissions budget based on domestic emissions reductions alone. Australia is the developed country with the highest per-capita emissions in the world, and has made sluggish progress to change that. Ongoing delays to cutting emissions domestically and weak federal policy and targets have made it ever harder because we are rapidly using up our fair share of the emissions budget (one could also argue that we have already eaten up that 'fair share' in the past), leaving future Australian generations with an increasingly difficult task.

Australia will have to rely on contributing more than others in ways other than domestic emissions reductions, and explaining to partners in the international community the value of these other contributions. One can be global finance, where Australia at last seems to be willing to make a large contribution to climate change mitigation in the South-west Pacific and Southeast Asia. Other important mechanisms include technology transfer and emissions trading under Article 6. So would timely, secure supply of cheap, abundant, green energy and energy transition minerals to the world, providing a pathway for other countries to reduce dependence on use of fossil carbon. This pathway would be married with strong economic self-interest – making it potentially palatable across the political spectrum. Another way in which Australia could take responsibility for imposing more than a 'fair' share of climate change on the world would be a strong program in sustainable negative emissions. Australia has the geological reservoirs and land to pioneer sustainable biomass and carbon capture and sequestration projects (BECCS).

However, by far the most important contribution would be to stop being a laggard. The world is turning its eyes towards Australia after repeated wildfires and floods. The world could instead be watching Australia for its ability to turn its economy away from the old fossil-fuel-growth model into a renewable energy powerhouse.

Appendix 2.1 Derivation of global and Australian emissions budgets

This appendix outlines an approach to deriving global and country-specific emissions budgets from the carbon budgets contained in the IPCC Sixth Assessment report.

Global remaining carbon budget

We start from the global remaining carbon budgets reported by the IPCC (Sixth Assessment Working Group I report Table 5.8). These budgets apply from the start of 2020 and are quantified on the basis of warming relative to 1850–1900. We convert these into global remaining carbon budgets from 2013[42] and for temperatures relative to pre-industrial times. Our reduction of the budgets reflects the fact that the Paris Agreement clearly states that the targets are relative to pre-industrial. We make this reduction on the basis that the IPCC's Sixth Assessment Report (Cross-chapter Box 1.2) assesses the anthropogenic warming between

Table 2.2 The global remaining carbon budgets, adjusted for pre-industrial reference levels.

Temperature level and likelihood of staying below	Global remaining carbon budget from 2020 (GtCO$_2$)	Enlarging budget to account for global emissions between 2013 and 2020 (GtCO$_2$)	Reducing the carbon budget to make it relative to true pre-industrial (1750), rather than early pre-industrial (1850–1900, GtCO$_2$)	Global remaining carbon budget from 2013 relative to pre-industrial (GtCO$_2$)
<1.5°C with 50%	500	277	−150	627
<1.6°C with 50%	650	277	−150	777
<2.0°C with 83%	900	277	−150	1027

Figure 2.10 The translation from cumulative CO_2 emissions to cumulative GHG emissions up to 2050 – on the basis of the latest IPCC Sixth Assessment Working Group III scenario database of approximately 1200 multi-gas scenarios. There is an almost linear relationship between the two quantities, allowing us to translate the one into the other and address the policy need for a quantity expressed in terms of the GHG emission baskets considered under the Paris Agreement.

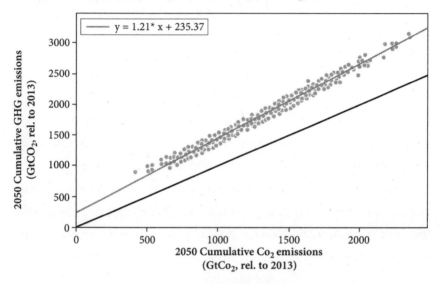

1850–1900 and the period around 1750 (before industrialisation, i.e. true pre-industrial) to be 0.1°C (with a likely range of 0.0°C to 0.2°C).

Emissions budget

The next step is to convert the global remaining carbon budget, which applies to CO_2 only, into a global remaining emissions budget, which can be applied to emissions of all greenhouse gases.

To do this conversion, we use the relationship between cumulative CO_2 emissions and cumulative greenhouse gas emissions derived from the IPCC Sixth Assessment Working Group III database. We calculate cumulative emissions between 2013 and 2050, as the shares of the global budget, calculated by the CCA (2014), were derived on the basis of cumulative emissions between 2013 and 2050. These shares can also be applied to the share of emissions between 2013 and net zero, as strong

Table 2.3 The conversion step from the global remaining carbon budget to cumulative GHG emissions up to 2050

Temperature level and likelihood of staying below	Global remaining carbon budget from 2013 relative to pre-industrial $(GtCO_2)$	Additional non-CO_2 greenhouse gas emissions until peak warming $(GtCO_2eq)$	Global remaining emissions budget from 2013 relative to pre-industrial $(GtCO_2eq)$
<1.5°C with 50%	627	370	997
<1.6°C with 50%	777	402	1179
<2.0°C with 83%	1027	456	1483

mitigation scenarios typically reach global net zero around 2050. The conversion is based on a correlation from cost-optimal emissions reduction scenarios, not any physical law; hence, it is not exact and comes with an uncertainty of around 100 $GtCO_2$-eq.

Differences in LULUCF accounting

Grassi et al.[43] identified a difference in the way emissions from land use, land-use change and forestry are accounted in country-reported emissions (those reported under the UNFCCC) and the emissions scenarios used by the IPCC (for example, those in the IPCC Sixth Assessment Working Group I database). In short, the main issue is that countries include the natural carbon cycle uptake due to CO_2 fertilisation in the areas that they declare as 'managed forests' in their reporting. While these forests are managed, the uptake is not anthropogenically induced: that is, it would occur even without management. The uptake is simply a consequence of us increasing atmospheric CO_2 concentrations, thereby slightly enhancing plant growth. Accounting for the inclusion of that indirect effect is as wrong as if somebody who drank four pints arguing that they should pay for three, given that they left the fluid equivalent of one pint in the bar's bathroom. (From our

Table 2.4 Adjusting the emission budget for the difference in IPCC and UNFCCC accounting approaches for land-use-related emissions

Temperature level and likelihood of staying below	Global remaining emissions budget from 2013 relative to pre-industrial ($GtCO_2$)	Adjustment to CO_2 part of emissions budget to account for different CO_2 sink accounting in IPCC methodology for national inventories and IPCC methodology for remaining carbon budget (see Grassi et al., $GtCO_2$)	Global remaining emissions budget from 2013 relative to pre-industrial after LULUCF adjustment ($GtCO_2$)
<1.5°C with 50%	997	−94	903
<1.6°C with 50%	1179	−117	1063
<2.0°C with 83%	1483	−154	1329

carbon emissions, the carbon cycle redistributes around a quarter to the trees on land, a quarter to the oceans and leaves, so far, half of our historical emissions in the atmosphere). This reporting practice by countries takes credit for the natural carbon cycle uptake and is in contrast to emissions budgets reported by the IPCC and scenarios used by the IPCC, where only directly human-induced emissions are considered.

We therefore reduce the available budget, ensuring that the targets we set are compatible with Australia's reporting under the UNFCCC. Without this step, the budgets given to governments would be too large – if they apply their reported emissions – and would not be in line with the relevant temperature limits.

The adjustment based on Grassi et al. (2021) is made only to the CO_2 part of the emissions budget. In line with their research, we reduce the CO_2 part of the budget by 15 per cent.

Table 2.5 Adjusting the remaining cumulative GHG emission budget for emissions from international aviation and bunker emissions

Temperature level and likelihood of staying below:	Global remaining emissions budget from 2013 relative to pre-industrial after LULUCF adjustment ($GtCO_2$)	Removal of emissions from international shipping and aviation ($GtCO_2$)	Global remaining emissions budget from 2013 relative to pre-industrial after LULUCF adjustment and international shipping and aviation are removed ($GtCO_2$)
1.5°C with 50%	903	−50	853
<1.6°C with 50%	1063	−50	1013
<2.0°C with 83%	1329	−50	1279

Removing international shipping and aviation emissions

Australia's share of the international emissions budget was calculated after the removal of emissions due to international aviation and shipping (CCA, 2014). Here we calculate international aviation and shipping emissions based on CMIP6 SSP scenarios (which have the required level of sectoral detail) approximately in line with the Paris Agreement, specifically the SSP1–1.9 and SSP1–2.6 scenarios.

Downscaling the global budget to a national budget

The final step is to calculate Australia's national emissions budget. This is done by downscaling the global budget using a specified proportion that reflects Australia's fair share using different equity approaches, as shown in Table 2.1 (page 140).

3

THE NET-ZERO OPPORTUNITY FOR AUSTRALIAN MINERALS

Mike Sandiford

The net-zero carbon economy requires decarbonisation of all sectors of the economy, especially those which have traditionally been served by fossil fuels, such as electricity, transport and mineral processing. That translates to a huge increase in demand for magnets and batteries, and the many other technologies that provide the essential building blocks that will allow for the 'electrification of nearly everything'. In turn, that creates a huge increase in demand for many relatively scarce metals such as lithium, cobalt, neodymium, praseodymium and vanadium, as well as the more abundant metals such as iron, aluminium, copper and nickel that are necessary for the structural framework and much of the communication and transport infrastructure that has underpinned the development of the modern global economy. For rapid decarbonisation, extraordinary increases in supply of the critical energy transition metals will be necessary, both in terms of rates of growth and in magnitude.

For example, electric cars contain up to five times as much of these rarer, energy transition metals as conventional cars do.[1] Demand for lithium, nickel, manganese and a raft of scarcer metals such as cobalt, vanadium and some of the rare earth elements (REE) such as neodymium, will rise many-fold in a net-zero-carbon world. The International Energy Agency estimates in its net-zero pathway scenario that the total market for copper, cobalt, manganese and various rare earth elements will need to grow by almost 700 per cent between 2020 and 2030.[2]

But it is not just the new rare metals for which demand will increase. As the most copper- and steel-intensive form of energy production, and as a mainstay of a fully renewable energy system, wind power will drive high demand not only for the REEs but also for traditional industrial metals steel and copper. Each new megawatt of wind power capacity absorbs between 120 and 180 tonnes of steel and between 5 and 14 tonnes of copper. The expected deployment of 650 gigawatts of new onshore wind capacity and 130 gigawatts of offshore wind capacity by 2028 will see demand for copper rise to 5.5 million tonnes per annum,[3] an increase equivalent to about 25 per cent of current global production.

Deep decarbonisation raises a suite of urgent questions[4] such as 'where and how are we going to source them?', 'what will be the impacts of price shocks that will accompany inevitable demand-supply imbalance', 'what geopolitical consequences will flow from the winners and losers on the new chessboard of rare metal supply? and 'what are the economic, social, and environmental costs for securing supply chains?'.

A zero-carbon economy will afford significant economic and strategic advantage to countries with significant geological endowments of critical metallic minerals, especially if they have globally competitive

Table 3.1 US Geological Survey (USGS) list of critical elements, ordered by crustal scarcity (relative to iron), with Australian production and reserves by global rank and main uses. Light grey signifies current Australian production. White signifies no known significant Australian reserve. Dark grey signifies known Australian reserves with no current or only very limited production (for example, Australia has vast silicon potential but only very limited current production at SIMOCA's facility at Kemerton in Western Australia) but potential for globally significant future production.

USGS proposed list of critical elements; Australian production and reserve rankings

Element		Scarcity	Prod.	Res.	Use
Silicon		0.2	–	–	alloys, semi–conductors
Aluminum		0.684	1st	1st	almost all sectors of the economy
Iron		1	1st	1st	steel
Magnesium		2.42	10th	4th	alloy, for reducing metals
Titanium		9.96	3rd	2nd	white pigment or metal alloys
Hydrogen		40.2	–	–	energy carrier, fertiliser
Manganese		59.3	2nd	2nd	steelmaking, batteries
Zirconium		341	1st	1st	high–temperature ceramics, corrosion–resistant alloys.
Vanadium		469	–	3rd	alloying agent for iron, steel, batteries, batteries
Chromium		552	–	–	stainless steel, other alloys
Rubidium		626	–	–	research, development electronics
Nickel		670	6th	2nd	stainless steel, superalloys, rechargeable batteries
Zinc		804	2nd	1st	metallurgy to produce galvanized steel
Cerium*	REE	847	4th	6th	catalytic converters, ceramics, glass, metallurgy, polishing compounds
Copper		938	6th	3rd	almost all sectors of the economy
Neodymium*	REE	1,360	4th	6th	permanent magnets, rubber catalysts, medical and industrial lasers
Yttrium		1,710	–	–	ceramic, catalysts, lasers, metallurgy, phosphors
Cobalt		2,250	3rd	2nd	rechargeable batteries, superalloys
Scandium		2,560	–	–	alloys, ceramics, fuel cells
Lithium		2,820	1st	2nd	rechargeable batteries
Niobium		2,820	–	–	steel, superalloys
Gallium		2,960	–	–	integrated circuits, optical devices like LEDs
Praseodymium*	REE	6,120	4th	6th	permanent magnets, batteries, aerospace alloys, ceramics, colorants
Samarium*	REE	7,990	4th	6th	permanent magnets, as an absorber nuclear reactors, in cancer treatments
Gadolinium*	REE	9,080	4th	6th	medical imaging, permanent magnets, steelmaking
Dysprosium*	REE	10,800	4th	6th	permanent magnets, datstorage devices, lasers
Erbium*	REE	16,100	4th	6th	fiber optics, optical amplifiers, lasers, glass colorants
Ytterbium*	REE	17,600	4th	6th	catalysts, scintillometers, lasers, metallurgy
Cesium		18,800	–	–	research, development
Hafnium		18,800	–	–	nuclear control rods, alloys, high–temperature ceramics
Beryllium		20,100	–	–	alloying agent aerospace, defense industries
Tin		24,500	8th	3rd	protective coatings, alloys for steel
Europium*	REE	28,200	4th	6th	phosphors, nuclear control rods
Tantalum#		28,200	7th	1st	electronic components, mostly capacitors, superalloys
Arsenic		31,300	–	–	semi–conductors
Germanium		37,500	–	–	fiber optics, night vision applications
Holmium*	REE	43,300	4th	6th	permanent magnets, nuclear control rods, lasers
Terbium*	REE	46,900	4th	6th	permanent magnets, fiber optics, lasers, solid–state devices
Lutetium*	REE	70,400	4th	6th	scintillators for medical imaging, electronics, some cancer therapies
Thulium*	REE	108,000	4th	6th	metal alloys, lasers
Indium		225,000	–	–	liquid crystal display screens
Antimony		282,000	6th	5th	lead–acid batteries, flame retardants
Palladium		3,750,000	–	–	catalytic converters, as catalyst agent
Bismuth		6,620,000	–	–	medical, atomic research
Platinum		11,300,000	–	–	catalytic converters
Iridium		56,300,000	–	–	coating of anodes for electrochemical processes, as chemical catalyst
Rhodium		56,300,000	–	–	catalytic converters, electrical components, as catalyst
Ruthenium		56,300,000	–	–	catalysts, as well as electrical contacts, chip resistors computers
Tellurium		56,300,000	–	–	solar cells, thermoelectric devices, as alloying additive

renewable energy resources for processing minerals into metals. For countries with both, there will be distinctive and immense economic opportunity in developing zero-emission mineral supply chains that include both mining and metal processing. Australia is one such lucky country, and it seems the luckiest of all. Peter Farley estimates that domestic processing of iron ore into green iron could increase total value of Australian exports by $70–120 billion each year, depending on plausible assumptions about the costs of hydrogen production and electrolysers, as well as premiums and carbon prices.[5] That amounts to 15 to 25 per cent of 2021 total exports value, or 1.2 to 2.1 multiple on the 2021 rural goods export value. Added to the imperatives dictated by climate change, the production and refining of these scarce metals is increasingly associated with strategic geopolitical supply risks.

While future demand for the energy transition elements will rise rapidly, the specific demand for individual metals will depend on technological developments that have yet to play out. Manganese and vanadium are essential ingredients for contenders for large-scale 'flow' batteries, as is iron. Which technology wins out may turn on supply security as much as pricing or environmental risk.

Many of the scarcer metals essential to the energy transition are collectively referred as 'critical metals' or 'critical minerals'. The term 'critical' refers to both their use in essential technologies and their exposure to geopolitical threats posed by geographically limited supply chains. The United States Geological Survey (USGS) lists fifty elements in its 2021 critical mineral list, of which only twenty are critical to energy technologies.[6] The REEs are a particular concern because more than 80 per cent of the global supply is currently sourced from China.

Global supply chains for critical minerals concerns also include environmental, social, ethical and pricing risks. Mineral processing can incur severe environmental risks. In countries with governments that are prepared to tolerate higher levels of environmental risk, or apply lax regulation, refined mineral product can be supplied at a significant cost advantage. Rare earth supply is largely controlled by China, not just because of rich primary resource endowments, but also because of tolerance of refining processes that are reportedly already risking significant environmental damage.[7] Around 60 per cent of the global supply of cobalt is sourced from poorly regulated artisanal mining in the Democratic Republic of Congo, where the safety of miners is largely unregulated.[8]

Recently, the term 'critical' has gained currency in reference to the essential role of these rare metals in the energy transition. However, as noted earlier, the energy transition is equally dependent on the more abundant metals such as copper, manganese and iron. It is better to use a more inclusive term to summarise both the challenge and the opportunity. Here I use 'energy transition metals'.[9]

A central concern for energy transition metals is whether supply can match demand derived from achievement of net-zero carbon consistent with climate objectives such as limiting warming to 1.5°C. The International Energy Agency (IEA)'s net-zero pathway requires year-on-year growth of about 50 per cent for key metals over the next decade. This is against a backdrop where primary discovery rates for large metal deposits have been declining for several decades. The search space for mineral explorers is becoming riskier and more expensive.

There are concerns about the scale of future recoverable resources for conventional metallic minerals such as copper, for which the fundamental geochemistry is now well understood. The fraction of the stock of recoverable copper already in use, or in wastes from which it will probably never be recovered, was estimated to be already at ~26 per cent as long ago as 2005.[10] While such estimates are highly uncertain, it points to the challenge in any massive ramp-up in future demand as implicit in net zero carbon. How do we ensure we find enough of the stuff to supply demand? The question of peak supply for metals has been a recurring theme for resource economists, for over a century in the case of copper.[11]

Many of the metals that are important for the energy transition are scarce and have few or no primary geochemical mineral enrichments. Such metals are referred to as 'companion' metals and are usually sourced using secondary recovery processes from ores that are mined for primary metals such as copper. For many companion metals, there are likely to be substantial stocks in existing mining wastes, such as smelter slags and tailings. For some companion metals, such as indium and scandium, the lack of fundamental geochemical knowledge about their distribution means assessments of resource availability are extremely uncertain.

Of great importance for the energy transition is the likely severe price impacts of inevitable supply-demand imbalance. If supply cannot meet demand at any time, then price of the scarce material will rise sharply, rationing limited supplies through the price mechanism. Some manufactures that depend on the input will curtail production, reducing demand and forcing price to lower levels. The associated price and supply instability can reduce investment both in the mineral supply and the minerals-using industries. Extreme volatility will slow movement along the decarbonisation pathway.[12] The International Monetary Fund (IMF) estimated that real prices of nickel, cobalt and lithium could rise several hundred per cent from 2020 levels on the IEA's net-zero emissions scenario,[13] with the total value of the metals production potentially increasing more than fourfold. In such a scenario, total value of these metals along with copper is anticipated to rival the 'value of oil production in a net-zero emissions scenario', potentially yielding extraordinary windfalls for producing and exporting nations. Of particular relevance for Australia is the IEA anticipation that, on its net-zero pathway, the total value of critical metals will exceed that of the global coal trade as early as 2030. Pricing concerns are already being realised, with scarcity pricing currently delivering significant windfall gains for resource producing countries. Across 2021, lithium, cobalt and neodymium prices rose by 525 per cent, 95 per cent and 75 per cent, respectively. Lithium prices doubled again in the first three months of 2022.

Even in very optimistic outlooks, with all new and proposed resource projects currently in the pipeline eventuating, raw material supply will likely limit future battery demand. Some forecasts have lithium carbonate supply falling short of currently indicated production by more than 300,000 tonnes by 2030,[14] equivalent in magnitude to the global production in 2020. Nickel sulphate may be in deficit by nearly 400,000 tonnes, cobalt by over 75,000 tonnes and flake graphite by nearly 2 million tonnes. If such supply constraints were to apply, they would severely limit battery production, possibly to no more than about 1 TWh per annum until well after 2025.

Supply concerns are by no means new. Long ago, noted resource geologist Brian Skinner raised concerns about the mineralogical barriers that may limit resource availability, especially for the scarcer 'companion' metals.[15] However, historically the supply of raw commodities has quickly followed rising demand and prices, keeping resource prices in check. This phenomenon was most famously evidenced in the Erhlich–Simon wager.[16] In 1980, Julian Simon challenged Paul Ehrlich to choose a basket of raw materials and a date more than a year away, wagering the inflation-adjusted prices would decrease, not increase. Ehrlich chose copper, chromium, nickel, tin and tungsten, and a ten-year period. When the wager settled in 1990, inflation adjusted prices for all commodities had decreased despite Ehrlich's anticipated demand pressures driven by rising population and an expected decline in natural resource inventories. The 'cornucopian' Simon won. Simon has been extremely critical of the notion of 'peak resources'.[17] For example, even though the idea of 'peak copper' has been persistent for almost

100 years, Simon points out that at least up until the late 1990s reserve growth outpaced demand, as evidenced by falling long-run price of copper. Simon reasons that rising prices provide impetus to seek technological improvements that when realised inevitably open substantial new resource opportunities. In recent times, this is most evident in the use of horizontal drilling and fracking in unconventional oil and gas plays that have largely put paid to the 'peak oil' concerns that were prevalent twenty years ago.

The transition to net-zero emissions will provide both challenges and opportunities for the Australian mining sector. As a resource-rich country, unique in its continental scale and geological diversity, Australia has advantages in its natural mineral resource endowments. Not only can Australia be a leading supplier for many of the raw commodities essential for the energy transition, but with its high-quality renewable resources it has natural advantage in a net-zero world as a potential home for energy-intensive mineral processing. Realising this potential will require an intensive search for new resources, and innovation in much less carbon-intensive approaches to mineral processing. This is set against a backdrop of a higher education system which is disinvesting in many of the most relevant disciplines, such as the geosciences. For several of the most important relevant disciplines, such as geophysics and metallurgy, the collapse in university training and research is especially alarming, with current trajectories headed for their extinction, just as the industry demand for such specialist training rises to unprecedented levels. Historically, Australia has been robust in delivering skilled workforces in fields such as geosciences and metallurgy. This can no longer be guaranteed.

The potential for a vastly expanded mining sector, inclusive of mineral processing, raises important issues for policy-makers across the spectrum of mineral resources development, with the need to balance environmental, social, security and economic concerns. How capital is distributed between resource recovery and mineral processing will require careful consideration, and appropriate policy development. The opportunity for mining and processing of critical minerals in Australia is geographically diverse and, managed appropriately, could provide a boon for regional and rural development.

Dig it up and ship it out

The Reserve Bank of Australia estimates mining makes up about 12 per cent of Australian industry output, and 60 per cent of export share.[18] Current production is dominated by iron ore. Annual Australian iron-ore exports now stand at over 900 million tonnes, accounting for about 36 per cent of total global demand and 56 per cent of global exports.[19] Australia is also the leading global supplier of bauxite (for aluminium), lithium and

Table 3.2 USGS production statistics for 2021.

Commodity	Australian production	Global production (%)	Global rank	Leading producer country, %
Lithium	40,000	49	1	Australia, 49
Iron ore	900,000,000	38	1	Australia, 38
Bauxite	110,000,000	30	1	Australia, 30
Manganese	3,300,000	18	2	South Africa, 28
Zinc	1,400,000	12	2	China, 35
Lead	480,000	11	2	China, 43
Gold	320	10	2	China, 12
Cobalt	5700	4	3	Congo, 68
Rare earths	17,000	7	4	China, 58
Nickel	170,000	7	6	Indonesia, 30
Copper	870,000	4	6	Chile, 28

Note: production units – metric tonnes.

rutile (for titanium); second in manganese, gold, zircon and lead; and third in uranium and copper. Australia also has significant reserves of rare earth elements, silicon, magnesium and vanadium, and prospects for much rarer critical metals such as indium and gallium.

The changing role of minerals in the Australian economy is highlighted by the monetary value of exported goods – or so-called goods credits (Table 3.2). Fifty years ago, metallic minerals accounted around 24 per cent of total Australian goods credits, compared with rural goods at 50 per cent. Currently the metallic mineral sector makes up around 44 per cent of the total, with rural goods at 12 per cent. Fossil fuels including coal and liquified natural gas (LNG) make up 25 per cent.

Table 3.3 USGS commodity reserve statistics.

Commodity	Australian reserves	Global reserve (%)	Global rank	Leading reserve country, %
Lead	36,000,000	41	1	Australia, 41
Zinc	25,000,000	37	1	Australia, 37
Iron ore	50,000,000,000	28	1	Australia, 28
Gold	10,000	19	1	Australia, 19
Bauxite	5,100,000,000	17	1	Australia, 17
Lithium	4,700,000	22	2	Chile, 44
Manganese	270,000,000	21	2	South Africa, 40
Nickel	20,000,000	21	2	Indonesia, 22
Cobalt	1,400,000	20	2	Congo, 20
Vanadium	4,000,000	18	3	China, 43
Copper	88,000,000	10	3	Chile, 23
Rare earths	4,100,000	3	6	China, 37

Note: Reserve units – metric tonnes. Note that comparison of reserves, are uncertain due to the different methodologies employed across countries. Australian reserve estimates are JORC compliant.

Since the turn of this century, there has been a dramatic rise of total mineral goods exports. This comprises both raw ore exports such as iron ore and refined products such as manganese metal, alumina and aluminium. To a large extent, the growth has been driven by burgeoning Chinese demand and is almost entirely attributable to primary ore. In contrast, there has been no significant growth in the value of refined metal exports.

Prior to 2000, the value of mineral ores export credits sat at about 1.8 times the refined metal production. This ratio is now at around fifteen and reflects the progressive shift in capital allocation in the resource sector almost exclusively to mine development, at the expense of mineral processing. The Australian resource sector is now essentially focused on the digging up of raw product, as perhaps exemplified by the industry 'dig it up and ship it out' mantra espoused most prominently by former BHP CEO Marius Kloppers.[20]

Figure 3.1 Ratio of value of exported Australian ore to exported Australian processed metals.

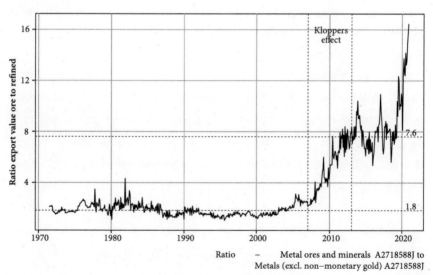

Ratio   –   Metal ores and minerals A2718588J to Metals (excl. non–monetary gold) A2718588J

Source: Data sourced from the Australian Bureau of Statistics.

Table 3.4 Goods export credits by sector as percentage of total.

	Rural	Fuels	Minerals	Other
1971–73	50	6	24	21
1999–2001	24	20	21	35
2019–21	12	25	44	19

Source: Data sourced from ABS 5368.0, TABLE 5[21], averaged over last two years.

Note: Fossil fuels aggregates 'Coal, coke and briquettes' and 'Other mineral fuels'. Minerals and metals aggregates 'Metal ores and minerals' and 'Metals (excl. non-monetary gold)'.

The 'dig it up and ship it out' approach has several important consequences. One is the relocation offshore of processing emissions, which are therefore not counted against Australia's national emissions accounts. This is significant. Scope 3 emissions from offshore processing to metallic iron of Australian iron ore exports in 2020 are more than 900 $MT\text{-}CO_2e$, or about 180 per cent of Australian domestic emissions.[22]

Outlook for energy transition metal resources

As a leading resource producer for a range of minerals, Australia has over 300 operating mines producing twenty-six mineral commodities.[23] In terms of tonnage, iron and aluminium ores are the largest primary metal ores mined in Australia. While not typically considered as 'critical minerals', they are critically important to the energy transition and will remain central to Australian mineral resource production. Here I explore the current status and outlook for recovery of several of the other main energy transition metals.

The energy transition metals represent a geochemically diverse array of metals, with prospective resources distributed widely across Australia. There are significant prospects in all states. The most prospective regions are in the traditional resource-rich metal provinces of the Yilgarn and Pilbara of Western Australia, the Gawler and Curna-mona in South Australia, Mount Isa in Queensland, and Broken Hill in New South Wales. In many instances the resource inventories for the critical minerals transition metals are poorly understood and there is much uncertainty. Resources include primary geological endow-ments, such as the lithium-bearing spodumene pegmatites in Western Australia, secondary enrichments where deep weathering has con-centrated metals in the near-surface laterites, such as applies to many existing Australian iron, bauxite and manganese deposits, which also host cobalt and scandium. In addition, there are significant resource endowments of metals such as cobalt in existing mine wastes, includ-ing smelter slags and mine tailings.

Lithium is the emblematic 'energy transition metal', being the essen-tial ingredient in the batteries that are currently in greatest demand. It is sourced globally from both hard rock mineral and brine mining. Brine mining is dominated by Bolivia and Peru, in the high Andes. The USGS reports total lithium 2020 production at 82,000 tonnes. The IEA's zero-carbon pathway anticipates lithium demand will grow thirty-fold to 2030 and more than 100-fold by 2050.

Australia is the leading lithium supplier, accounting for around 40 per cent of global supply. All Australian production comes from lithium-bearing pegmatites, with the principal ore-bearing mineral being spodumene. Current supply is from Western Australia. New mines are under active development in both Western Australia and the Northern Territory, which will see significant expansion of produc-tion in 2022. The lithium content of the Australian reserve (4,700,000 tonnes) had a value of approximately $500 billion at 2020 prices. Aus-tralia has no significant identified lithium brine resources.

The REEs comprise a group of the fifteen lanthanides, along with scandium and yttrium. The most critical REEs for the energy transi-tion are neodymium, dysprosium, praseodymium and terbium. These

Australian critical element mineral resources

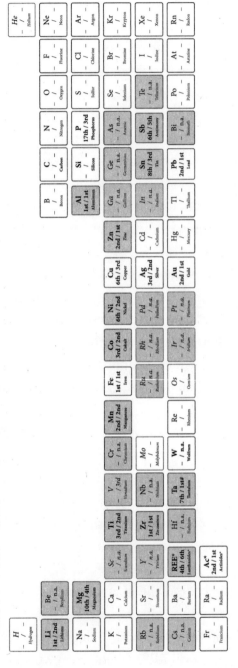

^1. Uranium rankings from from World Nuclear Acssociaton – https://www.world–nuclear.org/information–library/nuclear–fuel–cycle/mining–of–uranium/world–uranium–mining–production.aspx
^2. USGS reports lanthanide production/reserve data as bulk rare earths, rather than by individual element
critical element reserve data not available for Australia (n.a.) or poorly known/unknown for other leading producing countries (#)

Figure 3.2. 'Periodic table' of Australian metal production and reserves by global rank. USGS listed critical metals in shaded grey. Australian metals production/reserves by global rank in bold. Australian significant reserves by global rank but no production in italic.

are essential ingredients in permanent magnets that are vital for wind turbines and electric vehicles. Rare earth production is dominated by China, with Australia responsible for about 7 per cent of global production. Two leading Australian rare earth miners have plans for the development of mineral processing facilities in Australia. Iluka has committed to building a refinery at Eneabba in Western Australia, with agreement on a risk-sharing arrangement with the Australian federal government, including through a non-recourse loan under the government's $2-billion Critical Minerals Facility, administered by Export Finance.[24] The Lynas Mt Weld mine in Western Australia is one of the world's premier rare earths deposits. Currently Lynas operates the world's largest single rare earth processing plant in Malaysia, and it has announced plans to build a $500-million refinery at Kalgoorlie.[25]

Manganese is used primarily as a steel additive, and increasingly for batteries. Australia is the second-largest ore producer at 18 per cent global production and 21 per cent global reserve, with the majority of ore sourced from Groote Eylandt in the Northern Territory. About 10 per cent of the ore is processed domestically in Tasmania, with the remaining 90 per cent (approximately) shipped overseas. The long-term outlook for manganese production is unclear, with the current South32 mining operations on Groote Eylandt set for closure before 2030.[26] Most Australian manganese prospects are secondary enrichments in ancient, deeply weathered zones above manganese-rich host rocks. Deeply weathered metal-rich laterites are relatively abundant across northern Australia, though manganese concentrations are typically sub-economic.

Global cobalt production is currently dominated by the Democratic Republic of Congo (around 73 per cent). Australia accounts for about 5 per cent. Due to its low concentration in ores, cobalt is typically mined as a companion metal as a byproduct of extraction of copper, nickel and arsenic.[27] At current prices the value of cobalt in Australian

listed reserves is $114 billion. Given the significant historic produc-
tion of copper in Australia, there is significant potential for cobalt in
mine wastes, including smelter slag waste such as Mount Isa. Economic
cobalt prospects are probable in nickel-bearing laterites.

Vanadium usually constitutes less than 2 per cent of the host rock.
Because vanadium is typically recovered as a byproduct or co-product,
demonstrated world resources of the element are not fully indicative
of available supplies. At 60 per cent, China is the world's major supplier, followed by Russia (17 per cent) and South Africa (7 per cent).
In 2019, global production was 73,000 tonnes. Australia has significant demonstrated vanadium reserves, with mines in development in
Queensland and several deposits in Western Australia in pre-feasibility.
The Queensland resource is hosted by the Toolebuc Formation, an oil
shale in the Eromanga Basin outcropping extensively in the Julia Creek
region, as well as near Barcaldine. The vanadium content of the indicative Australian reserve of about 4 million tonnes has a notional value of
approximately $125 billion at 2020 prices. The Saint Elmo mine, which
commenced construction in 2022, is initially forecast to produce up to
5000 tonnes of vanadium pentoxide per annum (approximately 2800
tonnes of vanadium).[28] The Vecco group has advanced plans for mining at Debella and plans to construct and operate the first vanadium
battery electrolyte manufacturing plant in Australia. Australian Vanadium Limited has an advanced project near Geraldton in the mid-west
of Western Australia. Both mines are located near world-class renewable energy resources, which will facilitate low-emission processing.

The future of Australian mineral resource processing

Currently, the carbon emissions embodied in processing Australian
ore exceed 1 billion tonnes of CO_2-equivalent per annum, constitut-
ing many times Australia's own emissions. Almost all processing of
Australian ore is now undertaken offshore. In a future zero-carbon

world, global metal-refining capability will have to be totally rebuilt to address emissions intensity, energy supply cost changes and global supply chain issues. This will provide opportunity for countries such as Australia which have natural comparative advantage based on cheap renewable energy generation.

In addition to the natural advantages afforded by cheap renewable energy production and natural metal endowments, emerging geopolitical risks in global supply chains are one of the key defining issues for supply of many of the critical energy transition metals. There are therefore security reasons for developing an indigenous processing capacity. Battery supply chains are atypical in that they require much higher-purity metal inputs than for other uses. Consequently, there is need for very careful attention to the specific geochemistry of the primary ores and their metallurgy, something for which Australia's diverse natural endowments should provide competitive advantage.

In terms of conventional metals, such as iron, the principal source of emissions relate to the use of coal as the source of both energy for smelting (mainly heating) and as the reductant for liberating the metal from the oxidised primary ore. Decarbonising the metal refining process will require new technology based on non-fossil sources of reduction and energy. In the future zero-carbon world, the reductant of choice will be green hydrogen, and for energy it will be the cheap renewables, solar and wind.

The key challenge for renewable energy sources for metal refining relates to the need for relatively constant thermal loads. Thermal processing is essential to most current metal refining and is inherently inflexible. This is most notorious for aluminium processing, which is often jokingly referred to as 'congealed electricity'. In the main Hall-Héroult process, alumina is reduced to liquid aluminium metal at temperatures of about 950°C in 'pots'. Producing a tonne of aluminium metal requires roughly the same amount of electricity as is used by an average home in a year.

Consequently, energy consumption typically accounts for almost half the cost of aluminium refining.

While thermal processing can be varied, there are limits to the extent. For conventional aluminum processing those limits are likely to be in the range 10 to 20 per cent of total energy demand. For renewable energy supply, this adds substantially to the cost. New technologies such as EnPOT allow for up to 30 per cent intra-day variability in energy demand, allowing more scope for smelters to arbitrage energy market opportunities. Providing load 24/7 across all seasons within a limited range of variability will require a mix of strategies. In order to secure high returns on mineral processing capital investments, there will need to be significant energy storage, in chemical batteries, gravity systems and/or hydrogen, all of which add significant cost. The need for storage can be ameliorated to some degree by geographically dispersed power generation, provided transmission capacity is available. A major source of flexibility is the use of excess electricity at times of high renewable generation for electrolytic hydrogen production and the final stage of steelmaking using electric arc technology.

Emerging policy issues

Australia has very large reserves of several metals that underpin the modern global economy, including iron and aluminum. As such, Australia will remain a leading global commodity supplier for these commodities. Australia is already a leading supplier of lithium and manganese, which are essential to some existing and emerging energy storage technologies. The rare earth resource in Australia is huge, but its current supply is relatively small. Cobalt, vanadium and other, rarer critical metals are also relatively abundant, and the mining of these will help diversify the resource sector in terms of both commodities and geographical dispersion, and that should bring new opportunities to rural and remote regions.

In a zero-carbon world in which primary energy is sourced mainly from renewable resources, Australia will have significant competitive advantage as a potential home for metal refining and processing. Consideration of future large power demand centres associated with mineral processing will be important to planning a reliable, least-cost, national renewable energy system. Developing the local mineral-refining capacity will also help mitigate global supply risks, which are emerging for many of the critical metals, such as the REEs. With significant opportunity for the recovery of important companion metals such as cobalt by reprocessing of existing mine wastes, there is an opportunity for remediation of past mining activity.

Despite these advantages there will be challenges in realising the potential, particularly on pathways of aggressive decarbonisation. These include the global competition for capital and human resources.

While the cheap energy advantage afforded to Australian operations in a net-zero world is well recognised, we are not yet in that world. In the Australian energy markets, the marginal cost of electricity is historically set by gas pricing. Since the opening-up of the eastern Australian gas market to international gas pricing in 2015, domestic electricity prices have been de facto fixed to global gas price fluctuations. Consequently, Australia is currently a relatively high-energy-cost country. In this environment, encouraging large capital allocations into energy-intensive mineral processing in anticipation of cheap energy emerging on a decadal timescale will necessarily require some significant transitional financial incentives.

The research challenges in realising the benefit of Australian critical mineral endowments across the full supply chain remain immense and will require dedicated funding. This is now recognised with the Department of Industry, Science, Energy and Resources establishing a Critical Minerals Facilitation Office, including a range of funding initiatives. This includes an initial funding package for research and development of $50 million directed to CSIRO and Geoscience Australia as well as a raft of other measures through programs such as the CRC and NCRIS.

It is important to realise that these funding arrangements build on a strong historical record of higher education investment in relevant disciplines. That is no longer guaranteed. With a strong history of resource development, Australia has had ready access to highly trained professionals in fields such as geophysics, geology, metallurgy and mining engineering, as well as to world-class university research. This is no longer the case. Training in geophysics and metallurgy is close to extinction in this country, and with it the university-related research in these fields in which Australia has been a world leader in the past. It is bewildering that in a country so economically dependent on its earth resources, its principal research agency, the Australian Research Council, now allocates an order of magnitude more funding for the study of the physics of remote stars than it does for the physics of our own planet. The global pandemic has been particularly hard on the earth resource disciplines. In the face of a federal government that has appeared to be deliberately degrading university education and research, the higher education sector has clearly failed to see long-term advantage in resource sector training and research. Universities have socialised the loss from international fee-paying students (mainly in disciplines such as business to and economics) across all disciplines, and with weak domestic student enrolments this has seen devastation for the geosciences. Not long ago the Research School of Earth Science at the ANU ranked in the top five international research institutes in the field. It has been savaged, with some of Australia's most celebrated scientists forced into early retirement. In a panicked reaction to the pandemic, geosciences at University of Melbourne suffered a similar fate as it was cut harder than any other discipline, despite negligible international student exposure. Internationally recognised geoscience departments at Macquarie and Newcastle Universities were effectively closed, and with a few notable exceptions most other schools were downsized in attempts to balance the loss of international fee-paying students.

It seems impossible to conceive how disciplines such as geophysics and metallurgy can survive in the Australian university system if current funding arrangements remain directed by the whims of eighteen-year-olds' enrolments and the ambitions of international students. Some of our largest resource companies, such as BHP, are alert to and alarmed by the critical nature of this issue and are offering special graduate student internship programs to entice students to engage with the challenge. But this begs the question: will there be anyone left to train them in fields such as geophysics and metallurgy? If Australia is to realise its potential as a Superpower by expanding its supply of metals for the global energy transition, a policy imperative for the federal government should be to redress the current higher education funding arrangements to secure critical mass in these disciplines so essential to Australia's future national benefit.

4

THE DECARBONISATION OF ELECTRICITY

Dylan McConnell

A combination of policy measures, falling technology costs and consumer preference is driving a profound transformation of Australia's electricity system. Electricity supply is moving from a centralised system built around fossil-fuel generators towards a system dominated by geographically dispersed renewable energy generation.

This chapter assesses Australia's progress towards zero-carbon electricity, in the context of the need to decarbonise the economy more broadly. The transformation of the electricity sector, alongside the 'electrification of everything' is the cornerstone of efforts to dramatically reduce economy-wide emissions. The chapter starts by reviewing progress to date, and the resulting dynamics in the electricity market. It then looks at the outlook for the sector in the context of Australia doing its fair share in the global effort to avoid the worst impacts of climate change. This includes deep decarbonisation and electrification of other sectors of the economy and the utilisation of large export opportunities. It concludes by looking at the challenges faced in movement to a system dominated by renewable energy, and the associated market dynamics.

The beginnings of a renewable revolution

Over the last decade, renewable energy has grown to contribute more than 30 per cent of supply in the country's two largest electricity networks, the National Electricity Market (NEM) and the South West Interconnected System (SWIS) in Western Australia. The vast majority of this new supply has been supported by policy measures, helped along by significant reduction in renewables costs.

The main driver of deployment of renewable energy has been the federally legislated Renewable Energy Target (RET). The RET was meant to reduce greenhouse gas emissions by expanding supply of renewable energy. The RET required consumers of electricity to purchase a defined amount of generation from accredited renewable energy suppliers. It was introduced with a small target (9.5 TWh per annum by 2010) by the Howard government in 2000. In 2009, the target for large-scale renewable generation was increased greatly (to 41 TWh of energy per annum in 2020). Each retailer and large-scale user was required to acquire its proportionate share. Eligible renewable energy generators were allocated certificates, which energy consumers were required to buy. This created a market for renewable energy certificates. In addition, from 2009, a scheme for small-scale renewable generation (less than 100 KW) provided a proportionately similar level of support that was not included in calculations of the 41 TWh target. At the time, the large- and small-scale schemes were expected to represent 20 per cent of national electricity generation in 2020.

When the policy was first introduced, total costs of large-scale renewable energy were higher than both wholesale electricity prices and new-entrant coal and gas generation. Revenue from sale of the certificates was intended to bridge the gap between the cost of renewable energy and the alternative conventional supply, and market logic would lead to development of the renewable energy projects with highest value. Good projects that produced energy valued highly by the electricity market would deliver the lowest-cost certificates. As a consequence, the initial phase of the policy saw large expansions of wind power in South Australia and Victoria. Wind power was considerably cheaper than solar at the time, and South Australia and Victoria had good-quality wind resources with access to transmission capacity. South Australia also had relatively high wholesale electricity prices, because of greater reliance on gas and lower-quality, higher-cost coal. Wind projects in South Australia were particularly valuable as they generated certificates at low cost.

There was strong growth in both grid-scale and small-scale renewable energy generation after the strengthening of the support in 2009 (see Figure 4.1). However, the Abbott government sought to remove or to weaken the RET. In 2014, it set up a review of the target.[1] The work of

Figure 4.1 Variable renewable energy generation (wind and solar) in the National Electricity Market between 2009 and 2021. From the beginning of 2018, the supply of variable renewable generation grew by approximately 7.6 TWh per annum.

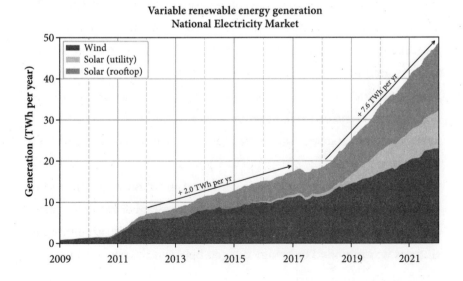

the panel and statements of government intention to truncate or abolish the target undermined confidence in the policy. But the Abbott government's intentions were blocked in the Senate. In 2015, a compromise with the ALP reduced the large-scale target from 41 TWh target to 33 TWh. After a brief pause, the rate of installation of renewable energy increased, to make up for lost time (see Figure 4.1). Renewable generation actually reached the reduced target before 2020. However, increasing demand for green energy by sub-national governments, companies and consumers, leading to voluntary surrender of certificates that do not contribute to the target, has maintained scarcity and positive value for the certificates, and maintained some continuing incentive for renewable energy investment.

This expansion occurred alongside significant reductions in the costs, most notably for solar photovoltaic (PV) technologies. This meant that the cost of renewables certificates was lower than it would have been. The increase in renewable generation, bid into the market at prices far below thermal power, lowered average wholesale electricity prices. This meant that total costs of wholesale power to retailers and users actually fell, even when the costs of certificate purchases were taken into account. The cost of renewable energy generation fell below that of new entrant conventional generation – and increasingly of the fuel costs of established thermal generation. Some renewable energy projects had become attractive without taking into account the value of renewable energy certificates.

The rise of the 'prosumer'

The renewable energy target and reduction in renewable energy costs also drove change at the individual household level. Many consumers chose their own 'behind the meter' energy solution. To the end of 2021, 3 million small-scale systems had been installed. Home batteries and electric vehicles are being chosen by a small but increasing number of households. These customers are both producers and consumers of energy (hence 'prosumers'), and a relatively new and significant addition to the electricity sector ecosystem.

The drivers of adoption of rooftop solar and other distributed energy choices include the desires to reduce household emission footprints, and to save money by reducing electricity bills and earn it from feeding electricity into the grid. Some of the monetary value is derived from reducing wholesale energy costs and some from reduced network charges. While environmental concerns and policy support remain important, the falling cost of solar in the midst of rising retail electricity prices supercharged the deployment of solar. Australia has the highest proportionate penetration in the world. The installed capacity of rooftop exceeds utility solar.

Within the federally legislated renewable energy target, small-scale renewable systems attract an upfront capital grant. For a typical system installed today, the upfront grant is worth about $3000, or 30 per cent of the capital cost. Unlike large-scale generation, this scheme is uncapped. The grant capitalises annual revenues for fifteen years, or the number of years from date of installation to 2030. As a result, it phases down through the 2020s to zero at the end of the decade. The scheme has enjoyed bipartisan support and has been only modestly altered since its inception.

The federal scheme was accompanied by high state incentives in the early years, varying across jurisdictions. These involved high rates of assistance in the early years and were subject to large and sometimes sudden change. They ceased to be significant in many states a few years ago. These changes at a sub-national level contributed to what has been dubbed the 'solar coaster', with the sector going through booms and busts. The rooftop solar industry nonetheless experienced record growth in recent years. Surges in wholesale energy costs in early 2022 have also seen an additional boom in sales inquiries. However, growth is also expected to face headwinds from broader supply chain issues.[2]

Wholesale market dynamics

Renewable energy has dramatically reshaped wholesale electricity market dynamics over the last decade, with complex effects on pricing.

Additional renewable energy supply places downward pressure on wholesale prices. Since renewable power generation has no fuel costs, the *short-run marginal* costs of renewable technologies such as solar PV and wind are negligible. Marginal costs are negative when the value of renewable certificates is taken into account. The low marginal price means that renewable energy is generally dispatched in preference to higher-marginal-cost generation like coal and gas, which must pay for fuel. Less conventional generation is required to meet demand – and to the extent thermal generation is required, a higher proportion can be drawn from sources with lower costs (for example, coal ahead of gas). This lowers the price for all power, since electricity markets operate as an auction within which the highest-cost generation supplied in each five-minute period sets the price for all.

Most distributed generation such as rooftop solar PV is not sold through the central dispatch system. It affects wholesale markets as an apparent reduction in demand. However, its impact on the wholesale spot price is similar to centrally dispatched renewable generation. Distributed PV lowers prices by subtracting from demand. The end results are similar: lower-cost generation is required to meet demand, and the wholesale price is lower.

This impact of additional renewable energy on the dispatch of electricity is now well documented and understood. In its 2021 update of retail price trends, the Australian Energy Market Commission pointed to the 'influx' of new generation driving wholesale, and ultimately retail, prices lower.[3] In a regular report on wholesale market dynamics, the Australian Competition and Consumer Commission noted that higher wind and solar generation had contributed to lower wholesale costs.[4] Indeed, even the 2014 review of the Renewable Energy Target, established to make a case for truncating or abolishing the RET, ultimately conceded that 'overall, the RET is exerting some downward pressure on wholesale electricity prices'.[5]

In markets with high proportions of renewable generation, it's now increasingly common for prices to drop to or below zero. Prices fall below zero since some generators (typically coal plants) are prepared to *pay* to generate for a period, rather than go through a costly shutdown cycle. During the fourth quarter of 2021, spot prices were negative 16.6 per cent of the time.[6] In the regions with the highest penetrations of renewable energy, South Australia had negative prices 28 per cent of the time and Victoria 24 per cent of the time.

South Australia exemplifies the impact of renewable energy on wholesale electricity prices. Historically, South Australian wholesale prices have been at the upper end of the regional price range. Factors

Figure 4.2 Volume weighted average wholesale prices since the beginning of the National Electricity Market to 2021–22, in real 2022 AUD. South Australia has moved from distinctively the highest-priced market region in the mainland NEM to the lowest or second-lowest as its penetration of renewable energy has increased. The impact of gas prices and tight supply–demand balance can clearly be seen in prices between 2016 and 2019.

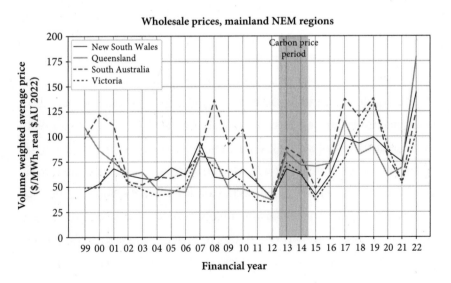

contributing to higher prices included greater reliance on gas generation (more expensive than coal), poor-quality and high-cost coal, relatively weak interconnection and limited competition.[7] Since the state's two brown coal generators closed in 2012 and 2016, renewable energy, including rooftop solar, has grown to contribute 66 per cent of the state's supply in 2021 and the reliance on gas has decreased. South Australia now enjoys the second-lowest wholesale prices in the mainland NEM (see Figure 4.2). The renewable energy share of electricity supply in South Australia is expected to reach 100 per cent on a net basis before the end of the decade, and then to keep rising as South Australia becomes a large net exporter. This will underpin relatively low wholesale power prices.

Wholesale prices were disrupted from early 2020 by the effects of COVID-19 on demand and then again more significantly by a combination of factors in 2022. Sharp lifts in coal and gas prices resulting from the conflict in Ukraine combined with significant coal plant outages to increase wholesale prices in the NEM to unprecedented levels. All else being equal, renewable supply will have kept prices lower than they otherwise would have been.[8]

Too good to be true?

At face value, this is a good news story. Renewable energy can reduce emissions while also reducing prices for everyone. The reality is more complex, as we have seen in part with wholesale pricing in 2022. The low prices and influx of new renewable generation also contributed to the earlier retirement of coal capacity. This has the temporary effect of reducing supply, which in turn lifts prices and results in higher-cost generation being required to meet demand. In the longer run, the supply shortfall is gradually reduced by expansion of renewable supply.

In a normally functioning wholesale market, prices and investment decisions are linked to the supply–demand balance. A tighter balance increases frequency of high-price scarcity events and elevates prices more generally. This encourages new supply capacity and corrects the balance. The new supply causes market prices to fall below long-run marginal costs, discouraging new generation.

Historically, this investment cycle was expected to be based on growing demand. Growing demand would increase prices, new supply would enter the market, prices would fall and the process would repeat. But declining or flat wholesale demand since 2008 – driven partly by rooftop solar generation and by increased efficiency in use of electricity – has changed the environment in which new renewable energy has been supplied. In the absence of growing demand, additional capacity reduces prices. Should prices fall low enough, this encourages retirement of capacity. The market tightens, price increases and new investment in generation capacity is encouraged again.

Coal-based generators are so large that the retirement of one can, in the short term, overwhelm the effects of incremental growth in renewable energy supply. The sudden exit of significant generation capacity can directly and dramatically alter prices. This dynamic can be illustrated with the closure of the Northern Power Station (520 MW) in South Australia in 2016, which was shortly followed by the closure of the larger Hazelwood Power Station (1600 MW) in Victoria in 2017. Alinta, the owners of the Northern Power Station, noted that the 'significant growth in renewable energy generation had the effect of causing a significant oversupply of power available to South Australia' and that there was no way they could remain profitable.[9] Similarly, the owners of Hazelwood pointed to 'difficult market conditions, with lower electricity prices and a surplus of electricity supply in Victoria State',[10] though they were also facing an expensive maintenance bill.[11] There were other factors at play – both plants were quite old – but clearly the impact of renewable energy on prices and volumes contributed to decisions on closure.

The Northern and Hazelwood closures occurred around the time that the beginnings of LNG exports from Gladstone lifted eastern Australian gas prices to the much higher international levels. Domestic gas prices increased threefold over a few years. The reduction in coal capacity, the associated decrease in competition and a rise in export prices combined to increase wholesale electricity prices to levels that hadn't been seen since the beginning of the NEM. This period can be clearly seen in Figure 4.2, between 2016 and 2019.[12] The high prices during this period, combined with renewable support policy, saw record renewable energy investment even with reduced renewable certificate prices.

The closure of the Liddell black-coal-fired plant in New South Wales provides a different example. This is also an ageing coal-fired power station that is approaching the end of its technical life. In 2015, the owners of the plant, AGL, announced that they would retire the plant in 2022, giving a full seven years' notice. AGL said at the time that extending its life would be expensive and unreliable and that 'energy output from Liddell is best replaced with lower emissions and more reliable generation'.[13] The long notice of closure was intended to provide sufficient time for the market to respond and prevent a supply shock. The notice would have contributed to responses from the private sector to commit to new supply (including from AGL itself). However, it also exposed both AGL and the system more broadly to an extended fear campaign centred around threats at first to reliability, and then to price. The federal government ultimately intervened by requiring the Commonwealth-owned corporation Snowy Hydro to build a new gas-fired power station as well (creating another source of market uncertainty for investment decisions). One Liddell unit closed in March 2022. In the second quarter of 2022, the closure does not appear to threaten reliability.[14] In this case, the early notice of reduction of capacity induced new supply, albeit from both private and state actors.

The increase in renewable generation and distributed energy resources continues to put pressure on the market share of incumbent generators. Origin Energy announced in early 2022 the retirement of its Eraring plant in New South Wales by 2025 (3.5 years from the announcement date). Eraring is the largest coal-fired plant in the country, with a registered capacity of 2880 MW that supplied approximately 20 per cent of the state's electricity in 2020/21. Origin Energy stated that the 'growth in renewable energy challenges Eraring's financial viability'. Origin's CEO noted that 'the economics of coal-fired power stations are being put under increasing, unsustainable pressure by cleaner and lower cost generation, including solar, wind and batteries'.[15] Interestingly, Eraring is not the oldest plant remaining in New South Wales, and is relatively flexible. Its vulnerability comes from coal supply issues, and having some of the most expensive coal costs in the NEM. Figure 4.3 shows how the generation profile of the Eraring plant has evolved over the past few years. The volume of generation has significantly declined, which exacerbates the effect of price declines on sales revenue and financial viability. The impact of solar on the profile of Eraring's output is particularly evident.

As discussed, the Hazelwood and Northern closures occurred just as the domestic gas market was opened up to international pricing. While the circumstances are different, the change in international commodity prices is yet again playing a big role in the near-term formation of domestic electricity prices in 2022. The conflict in Ukraine has introduced another price shock, just as these NSW plants begin their retirement. The closures, in combination with unplanned outages at ageing coal plants, are occurring in an environment of extremely high prices, for which the dominant driver has been large increases in international coal and gas prices flowing back into domestic markets.[16] Even though many coal plants in Australia would have contracted supplies for a period at lower prices (and some are not exposed to international

Figure 4.3 Generation profile for the 2880 MW Eraring black-coal-fired power station in New South Wales, the largest in the country. While not the oldest or most inflexible coal plant, the energy produced at Eraring is relatively expensive, and is displaced by renewable generation ahead of other coal generators.

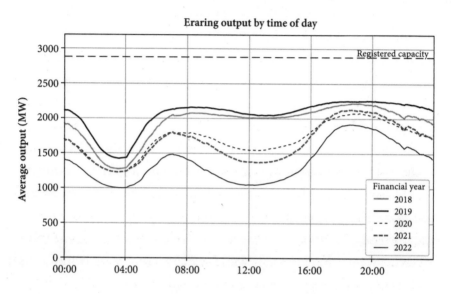

prices), wholesale prices in the NEM have increased with international coal and gas prices (Figure 4.2). Domestic gas prices have also lifted sharply under the influence of international market developments. The market dynamics are remarkably similar to those that played out when domestic gas prices increased from 2016. The forward prices for electricity are currently extraordinarily high, and well above historical averages for some regions. The states most reliant on black coal (New South Wales and Queensland) have the highest prices. There is considerable uncertainty about the duration of the conditions that have lifted international prices to such high levels, but their flow-on to domestic prices is exacerbated by the tightening of supply–demand balances resulting from coal closure and unscheduled outages.

While it is true that renewable energy reduces wholesale prices and emissions in the long term, the short-term dynamics are more complex.

As discussed, reduced prices and increased volumes are in part responsible for retirement of coal capacity. Sawtooth price movements can be expected: a downward trend from expansion of renewables supply, broken by an upward shift temporarily with each coal-generation closure. In the near term, the magnitude of each upward shift will be dictated by the marginal cost of fossil-fuel generation, which has proven to be volatile and expensive. In the longer term, the magnitude of the upward shift should decline over time as alternatives such as storage displace the role of fossil fuels.

Questions remain about the speed and political durability of this cycle. Supply shocks and price rises might invite further political intervention, undermining confidence in the market and introducing additional risks for prospective investors. The strength of incentives to decarbonise the economy, in the absence of policy to force renewable investment into the system, or withdraw coal capacity, may mean the cycle is not fast enough to meet emission objectives.

The next phase of the investment cycle?

It's clear that new renewable energy erodes the revenue of incumbent generation, by reducing both price and volume of sales. It erodes the price to renewable generators themselves even more, because it increases supply at exactly the time when renewable supply is greatest.

Prices fall to low or negative levels when renewable energy is abundant relative to demand at that time. Total electricity demand is greater in the daytime and early evening. Wholesale prices once were higher at these times. In recent years, the abundance of solar energy has overwhelmed high levels of daytime demand and caused prices to fall to low levels when the sun is shining. The average wholesale market value of renewable energy generation is generally at a discount to the load-weighted average price (LWAP) and the average price received by continuously operating thermal generators like those fuelled by coal. This remains true, even in the period of elevated pricing in 2022.

Figure 4.4 illustrates the relative value of different technologies in different market regions over time. Two points are worth noting. In regions with higher penetrations of renewable energy such as South Australia and Victoria, the discount on the market value is greater. And the discount for solar is substantially greater than the discount for wind generation. This reflects the extremely high correlation of solar production from different generators within a region. Note that utility solar competes with rooftop solar, so that supply of the latter depresses demand in the wholesale market and prices, while not formally competing in the wholesale market. The market value of solar has been particularly low in Victoria and South Australia, approaching $25/MWh in 2021. The value of renewable energy has since increased in line with the dramatic increases in wholesale prices. However, the value is still significantly discounted relative to overall prices.

This presents both opportunities and challenges. It opens up opportunities on the demand side, for users that can make use of cheap electricity when it is available. It also creates arbitrage opportunities and incentives for utility scale storage. In the last twelve months, the median daily price for two hours of arbitrage was more than $100/MWh in all regions (see Figure 4.5).

It may, however, present significant challenges for new investment in renewable capacity. Together with lower expectations of revenue from the RET, this is reflected in a slowdown in new investment in renewable generation. The Clean Energy Council reported that investment in renewable energy fell from $4.5 billion in 2020 to $3.7 billion in 2021.[17] The amount of new large-scale capacity committed fell from 3 GW in 2020 to just over 2 GW in 2021. The growth in wind power generation in particular has fallen from 20 per cent on a year on year basis to only 6 to 7 per cent. The dramatic increases in wholesale prices

Figure 4.4 Twelve-month moving average market value of different power sources (wind, utility solar and coal) compared with the load-weighted average price (LWAP). The influx of new renewable energy capacity has depressed the relative value of renewable generation. The value of solar is affected most, given the strong correlation of output with other utility solar and rooftop solar generation.

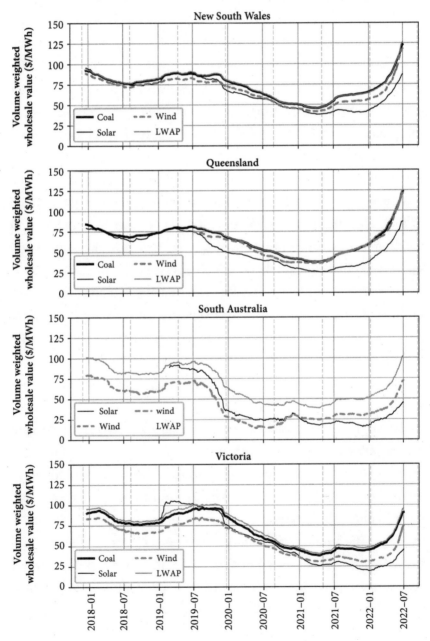

in 2022 may, however, see renewed interest in further investment in renewable capacity.

In the absence of a policy such as the Renewable Energy Target or carbon pricing to reflect the true costs of fossil fuels and constrain generation from coal, the electricity market is operating as one might expect. New supply can cause market prices to fall below the long-run marginal costs, discouraging new entry of generation. This phenomenon is behind the reality that the RET lowers average price of power even after consumers pay for certificates – at least up to some reasonably high proportion of renewables. However, low electricity market prices for renewable energy discourage investment in the energy transition and decelerates decarbonisation of the economy.

Market adjustments may restore incentives for new renewables investment over time. Market dynamics may accelerate the roll-out of battery storage or, on the demand side, energy-intensive industries may make use of periodically low prices. Investment in storage and flexible

Figure 4.5 Intra-day price spread for mainland NEM regions. The figure illustrates the rolling median price difference between the lowest and highest consecutive two hours period each day, over time. The influx of solar increased the intra-day price spread considerably after 2016.

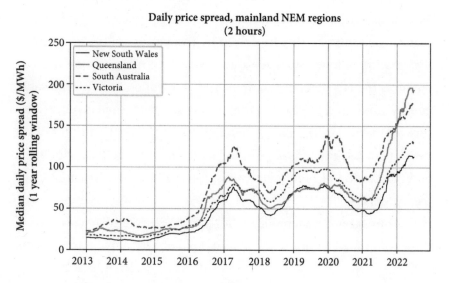

demand can in turn increase prices at times when renewable supply is abundant and reduce it at other times. However, in mid-2022, there is no 'engine', in Ross Garnaut's words,[18] such as a carbon price or RET driving this cycle, and investment had shown signs of slowing. This contradicts what is needed for decarbonisation.

Problems with connecting renewable generation to the grid in favourable wind and solar locations have also contributed to lower rates of investment. There is a mammoth task for public policy in planning and implementing the reshaping of the transmission grid so that it is more closely aligned with the requirements of an electricity system dominated by renewable energy.

Future expectations

System planning is critical to inform investment decisions and deliver a secure and reliable power system with zero net emissions. The importance of planning has long been recognised, with the AEMO playing the role of national transmission system planner since its establishment.[19] This includes reviewing and publishing advice on the development of the grid, to assist with planning and coordination.[20]

This role and responsibility were expanded following the Finkel Review in 2017, which recommended a stronger role for the AEMO in planning.[21] This was in part driven by recognition of the geographic distribution of low-cost renewable energy resources, and the need to facilitate the efficient development and connection of new renewable energy zones, some of which may be remote from the current transmission network. The recommendations suggested that the AEMO develop an integrated grid plan to facilitate the efficient development and connection of renewable energy zones across the National Electricity Market.

The Integrated System Plan (ISP) is the AEMO's implementation of these recommendations. The ISP is a whole-of-system plan, which aims to provide a roadmap for the efficient development of the NEM over the next twenty years, and beyond.[22] The plan looks at a range of plausible scenarios and determines development pathways that ensure the delivery of a secure and reliable energy system at lowest cost. The scenarios explore different levels of emissions from the electricity sector, by applying different carbon budget constraints. The plan is updated on a two-year cycle.[23] It involves significant consultation with stakeholders in the industry and more broadly. Major inputs into the process include technology costs and demand expectations, which reflect, among other things, changes in the degree of electrification of various economic activities.

The latest version of the plan explores four scenarios: slow change, progressive change, step-change and hydrogen superpower.[24] At a high level, the scenarios are briefly described as follows:

- **Slow change** – challenging economic environment
- **Progressive change** – pursuing an economy-wide net-zero-emissions 2050 target progressively, ratcheting up emissions reduction goals over time
- **Step-change** – rapid consumer-led transformation of the energy sector and coordinated economy-wide action
- **Hydrogen superpower** – strong global action and significant technological breakthroughs.

In the latest ISP, AEMO also surveyed market participants to form a view of which scenario is considered 'most likely'. This is used as a 'central' scenario for planning purposes, which has implications for assessing the economic case for different transmission options. The step-change scenario is currently considered 'most likely'. Unlike previous plans, there is no longer a scenario that strictly reflects current policies.

There are many differences between the scenarios, but two critical assumptions are worth drawing out here:[25] the extent and speed of decarbonisation and the level of electrification and export.

Figure 4.6 Growth in variable renewable energy generation projected in selected scenarios out to 2030, in the AEMO Integrated System Plan. The step-change scenario, which electricity system stakeholders currently consider to be the 'most likely', would see renewable energy growing at approximately twice the rate it has in recent years.

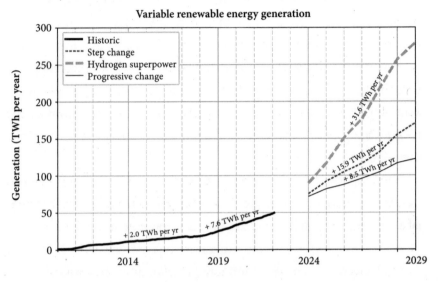

Figure 4.7 Growth in variable renewable energy generation projected in selected scenarios out to 2050, in the AEMO Integrated System Plan. The hydrogen superpower scenario envisages generation from variable renewable generation increasing almost thirty-fold from today's levels.

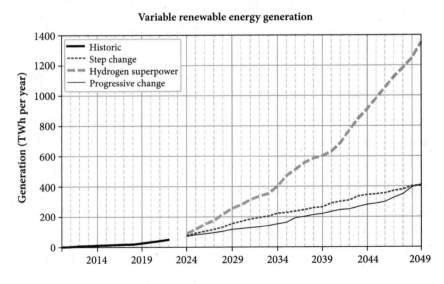

The step-change scenario has a carbon budget consistent with limiting temperature rise to 2°C. Note that this falls short of the 1.5°C that was favoured by the international community at Glasgow. This scenario has electricity sector emissions falling by 72 per cent of their 2005 levels by 2030 (and below 90 per cent by end of 2035), and results in all coal departing the system by the end of financial year 2042. The hydrogen superpower scenario has a budget consistent with limiting temperature rise to 1.5°C. This has electricity sector emissions falling by 90 per cent of their 2005 levels by 2030, and all coal leaving the system by the end of the 2030 financial year. The remaining scenarios have no binding exogenous carbon budget.

On electrification and export, the hydrogen superpower scenario has substantially greater demand, largely coming from the development of a hydrogen export industry, and to a lesser extent the development of 'green steel' (which involves iron ore being reduced by hydrogen through electrolysis using renewable energy, rather than coking coal). This scenario would see the annual export of 12.7 million tonnes of hydrogen and 50 million tonnes of green steel by 2050. The hydrogen superpower scenario would see electricity consumption increase to just on seven times the total volume generated in the NEM today. There are additional sensitivities that look at high levels of electrification (within a 1.5°C carbon budget) without the development of a substantial export industry. Scenarios with high electrification but without expansion of renewable exports see consumption rise by a factor of two.

Using the most up-to-date technology costs and projections, along with assumptions around emissions and electrification, the market operator determines the development of a 'least-cost' electricity system. Through this process, they explore so-called 'optimal development pathways', which map out the optimal timing, location and type of resource required to ensure secure and reliable operation of the power system for the particular scenarios. This includes investment in transmission as well as investment in new renewable generation and battery capacity. Note that this is a top-down centrally planned projection of possible futures, rather than a market modelling exercise based on estimation of wholesale prices and commercial responses by individual private firms.

Given the cost advantage of renewables, all scenarios see dramatic expansion of renewable generation over the planning horizon. The extent of electrification (and expansion of exports) and the carbon budget largely drive the extent and speed of that expansion. Figure 4.6 illustrates the expansion of renewable energy in the 'least cost' development pathway for three different scenarios out to 2030, while Figure 4.7 shows the same out to 2050.

The two scenarios with binding carbon budgets illustrate that the rate of roll-out of renewable energy (see Figure 4.1) needs to increase rapidly. Over the past four years, renewable energy has been added to the NEM at the rate of approximately 7.6 TWh per annum. The 'most likely' 2°C step-change scenario has renewable energy added to the grid at twice that rate in the medium term, while the hydrogen superpower scenario has it increasing at approximately four times that rate.

The hydrogen superpower requirement for new renewables generation reflects the substantial increase in electricity demand as well as a tighter carbon budget (Figure 4.7). Achieving the levels of renewable energy envisaged in the hydrogen superpower scenario requires adding the equivalent of all the renewable energy generation ever added to the NEM, every year, from now until 2050. This scenario would see an additional 245 GW of wind and almost 270 GW of utility solar.

The expansion of renewable energy and compliance with emissions constraints also implies acceleration of the retirement of ageing coal generation. The AEMO's ISP now expects 14 GW of coal to retire by 2030[26] and coal generation to fall 60 per cent by 2030 (see Figure 4.8). This implies a rate of reduction approximately three times faster than

Figure 4.8 Projected coal generation in the step-change and hydrogen superpower scenarios in the AEMO's Integrated System Plan. Electricity system stakeholders currently consider the step-change scenario to be the 'most likely'. The rate of decline in the step-change scenario is approximately three times that in the last fourteen years.

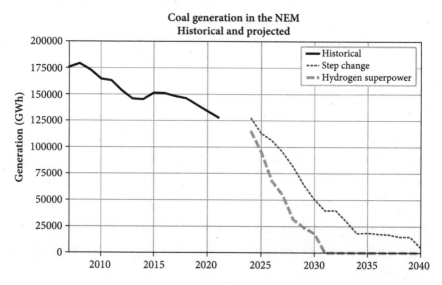

the rate it has fallen since its peak in 2008. The closure dates announced by market participants are well behind the rates implied in the ISP, even with recent announcement from AGL and Origin, bringing forward closure dates. A bid by Mike Cannon-Brookes to acquire AGL and a plan to bring forward closure dates would bring the coal closure more closely in line with the step-change scenario.[27] However, a considerable mismatch remains between market intentions and the closure schedule required in a 1.5°C carbon budget.

Perhaps the most important content for a network planning document is the amount, timing and location of transmission network investment. The plan identifies 10,000 km of new transmission to connect renewable energy zones and deliver renewable energy to consumers through the NEM.[28] The plan identified network investment of approximately $12.8 billion in today's value as 'actionable'.[29] For reference, the total regulated asset base of the transmission network in the NEM is approximately $22 billion in 2022 and includes about 40,000 km of transmission lines.

However, these actionable projects represent only a fraction of the total transmission identified in the plan, particularly for the more ambitious hydrogen superpower scenario. Of the thirty-nine renewable energy zones, only the New England REZ in New South Wales is currently considered an 'actionable' project, while the Orana REZ (also in New South Wales) is the only REZ considered to have been committed by the AEMO. The total amount of transmission investment is expected to be much higher than that. The ALP's current Rewiring the Nation policy is aimed at facilitating $78 billion in investment in generation and transmission over a decade, and is based on accelerating projects identified in the ISP. This policy is currently under development and centres on using $20 billion of concessional financing to 'rebuild' Australia's grid. This should ultimately help lower the cost of transmission projects, but questions remain about how projects are assessed in the current regulatory framework (and in a timely manner), and how the costs are ultimately recovered from energy consumers. There are also important questions about whether a larger departure from established approaches is required to underwrite emergence of the hydrogen superpower.

The uncertain road ahead

Australia is well placed to take advantage of the opportunities afforded by the energy transition. Over the past decade, we have seeded a successful renewable energy industry that could provide a useful launching pad from which to take advantage of the fall in technology costs, the country's world-class renewable energy resources and the growth in world demand for zero-emissions energy-intensive goods. The market operator has mapped out several scenarios that illustrate how this could unfold, ensuring the lowest-cost, reliable and secure system that also meets emission objectives.

However, there are some clear challenges ahead that mean that the realisation of some of these futures is not at all assured. Most significant is the discordance between current policy settings and the least cost pathways that are considered 'most likely'. The 'most likely' step-change scenario has renewable energy expanding at twice the rate over the past few years, and yet investment has been slowing. This points to the difference between the centrally planed least cost model and the market model alluded to earlier. Without policies driving more investment to support zero-emissions generation, or encouraging high-emissions plants to leave, the supply of renewable energy may be constrained by its own impact on prices. The recent political economy of energy policy in Australia makes deliberate government action to accelerate the retirement of coal unlikely. While these lower prices may reduce the viability of coal, and force its exit eventually, it is not clear that it would drive emissions reductions and renewable power expansion at the rate required.

This point is made by the market operator. Specifically, it has highlighted that in the case of coal closure, 'the carbon budget was the dominant driver for retirements in the Step Change and Hydrogen Superpower scenarios'.[30] It assumes that businesses and households act as if a carbon price or carbon constraint was driving behaviour. Needless to say, there is no binding carbon constraint or price in contemporary Australia. In response to the degree of electrification, the AEMO also noted that 'while these [changes] are likely over time, the pace so far has been relatively slow, and urgent action would be needed to put south-eastern regions on the Step Change path by next winter' (that is, the winter of 2022).[31] This indicates that the market operator judges that we are currently not on the 'most likely' pathway.

For renewable energy, the value of renewable energy certificates under the RET has been by far the most powerful engine driving renewable investment to this point. In the absence of this driver, the low prices for solar and wind power may not support investment at levels envisaged in the ISP roadmap. This is certainly true for the hydrogen superpower scenario. Perhaps state-based initiatives or voluntary action can fill the gap, but that remains to be seen. Whatever the case, some of these scenarios are currently based on an 'imaginary engine'.[32]

In addition, the uncoordinated withdrawal of such a substantial amount of capacity in such a short period of time could result in periods of tight supply, if left to the market. Even if driven by additional supply, temporary upticks can be expected, with a longer-term sawtooth-shaped reduction in prices. In a market like Australia's, with limited interconnection and a high degree of market concentration, this is likely to result in high and volatile prices. The facts that renewable energy is undeniably the lowest-cost source of new generation, and that the proportion of renewables in power supply will continue to increase, do not mean that electricity prices will necessarily remain low in the short to medium term, as the experience of 2022 has shown. Expansion of storage capacity and demand from energy-intensive industries that use power flexibly could help to create conditions for sufficiently rapid expansion of investment in renewables, but this will not happen at the required rate without supportive policy.

These developments require high investment in innovation. Public economics demonstrates that socially optimal levels of innovation do not emerge from unconstrained private decisions in competitive markets. Innovation generates benefits for the community beyond those that captured by the private investor. The necessary rate of innovation requires public financial support.

There are additional challenges with the deployment of transmission. Transmission is critical to developing the highly geographically distributed renewable energy resources. Renewable energy developers consistently say that transmission access is a major barrier to the roll-out of new renewable energy projects.[33] Transmission is not expanding at a rate or cost consistent with the scenarios envisaged in the integrated system plan. Arguments remain about who pays and how the projects should be regulated and approved. Some transmission projects are controversial in the local communities through which they traverse, and in some cases face influential local opposition.

Finally, there are challenges related to the electrification of everything and the development of energy-intensive export industries. As mentioned, the 'electrification of everything' is not progressing as expected in the step-change scenario. Current policies are insufficient to secure electrification of household and business activities now using gas, oil and coal, and the electrification of transport in Australia is much slower than in other developed countries. The development of export industries needs to be carefully coordinated with both the supply of renewable energy and transmission development. The addition of significant new load could help the roll-out of more renewable generation. Coordination aside, there is little prospect of renewable energy supply growing at four times its recent historical rate, or demand growing rapidly from electrification or development of export industries, within current policy settings. It's also worth noting that export rates considered here are for the scenario developed by the AEMO (seven times current demand) and there are more ambitious plans that would require an even greater level of policy support.[34]

Australia is blessed with renewable resources and has a world-class renewable industry. From these perspectives, we are well placed to take advantage of the energy transition and global decarbonisation. But good outcomes will not be achieved with an 'imaginary engine', without substantial policy support. Deliberate policy action is required for Australia to do its fair share in the global decarbonisation effort, and to utilise our exceptional economic opportunities.

5

HYDROGEN

Frank Jotzo

Hydrogen is an essential building block in the transition to a net-zero-emissions world economy. It can be produced from zero-emissions electricity and thereby can be used to store and ship clean energy, usually from renewables such as wind and solar. In this way it can replace fossil fuels in many applications that are otherwise difficult to decarbonise.

Hydrogen will be crucial for the carbon-neutral production of some energy-intensive goods that currently account for large amounts of greenhouse gas emissions, and which are difficult or impossible to convert to electricity – so called 'hard to abate' processes. It is an alternative to fossil carbon in reducing oxide ores to metals – for Australia, most importantly for reducing iron ore to iron in the first stage of steelmaking. It is the chemical foundation for producing many industrial products – for example, ammonia for nitrogenous fertilisers and explosives. It will help decarbonise heavy transport.

Australia is perfectly positioned to be a large producer and user of 'clean' hydrogen, including for energy-intensive commodities for export, because of Australia's practically unlimited potential to supply low-cost power from the sun and the wind. Rates of production of hydrogen from electricity through electrolysis can be varied considerably without technical difficulties, so its production fits neatly into use of intermittent renewable energy such as wind and solar.

In a decarbonising world economy, large-scale plants in remote areas of Australia could produce hydrogen for energy export to energy-importing countries that are constrained in their renewable energy potential relative to population and industrial production. These include countries in northern Europe and East Asia. Shipping hydrogen is costly, so exports on tankers will be economical only where the options to produce hydrogen locally are constrained. This is the case, for example, in Germany, which is pushing to shed its dependence on Russian gas while moving towards a low-emissions economy and keeping its energy-intensive industries going. Hydrogen trade opportunities are becoming more realistic, and from European importers' perspective more urgent, as a result of Russia's war in Ukraine.

An even larger prize for Australia than shipping hydrogen as a fuel would be new resource-processing industries in Australia that use hydrogen as an energy input or reductant. Potentially the biggest industry that could emerge is iron and steel: in a decarbonised world economy, primary steel would no longer use the traditional steelmaking process that relies on coking coal as its reductant but new methods that rely on hydrogen and electricity. For Australia, as the world's largest iron ore producer and exporter, this opens a tantalising possibility of processing some of the ore into iron – or even into steel through the use of more renewable electricity. This would mean potentially extremely large value added in the commodity supply chain, and export revenue. Fertiliser production is another potential hydrogen-based export industry.

Whether large-scale zero-carbon hydrogen-based industrial exports become a reality for Australia depends on many factors, from technological aspects to relative costs in different locations, global supply chain considerations and geopolitics. Where large-scale hydrogen production becomes a reality, it will be crucial to design and implement these projects in ways that accommodate Indigenous communities' needs and preferences, that minimise adverse local environmental impacts and that maximise gain to the broader community, including through taxation revenue.

Hydrogen will also have a role in Australia's domestic energy use. It is well suited to the production of various commodities for the domestic market, for heavy transport, and to augment supply in some remote applications. Most of the existing pilot projects focus on blending small amounts of hydrogen into gas networks or are hydrogen refuelling stations, but these are not the kinds of uses that will be transformative economically or in the climate transition.

Uses, attractions and difficulties of hydrogen

Hydrogen's fundamental advantages include that it can be turned into usable energy without creating any pollution at the point of use, and that it can be produced by splitting water into oxygen and hydrogen with the input of electricity, in a process called electrolysis. The only global warming impact of hydrogen use is if it leaks into the atmosphere. Hydrogen production can be clean depending on the energy sources and process used.

If the electricity is generated from renewable energy, the resulting 'green' hydrogen is a zero-carbon fuel that can be transported and stored in large volumes and for long periods of time. Zero-emissions hydrogen can also be produced using nuclear power where this is feasible and affordable. The main production method to date is by processing gas or coal, which leaves residual carbon dioxide emissions (see further below).

Hydrogen is versatile. Its greatest uses today are as a feedstock in chemical industrial processes, mostly to make ammonia, including for fertiliser production; in oil-refining; and in some other industrial uses.[1]

It can be used as a reduction agent in iron- and steelmaking, in processes that replace coking coal with hydrogen and electricity. Using green hydrogen in steel production could result in a meaningful reduction in future global greenhouse gas emissions; this could become one of the main uses of hydrogen in the long term – and a major industrial opportunity for Australia.

Hydrogen can also be burned for heat and in turbines, for power generation or thrust. It can be used in electricity-producing fuel cells to produce electricity cleanly at any scale, powering anything from small telecommunications equipment to ships, although fuel cells to date are very uncommon.

The benefits of hydrogen as an energy carrier and storage medium have been known for a long time and the advent of a global 'hydrogen economy' has been predicted many times since the 1970s. Yet today, hydrogen still accounts for only a tiny share of global energy use. It is mostly used in the chemical industry, including refining and production of ammonia.

Given its desirable properties, why is the world not already using very large amounts of hydrogen, and what might be different this time?

The main obstacle to hydrogen as a mainstream fuel is that much energy gets wasted in the production and compression and storage of hydrogen, as well as its conversion back to energy. The round trip from electricity to compressed hydrogen and back to electricity via a fuel cell has an efficiency cost of around 70 per cent of the total energy, so only about 30 per cent remains as usable electricity.[2] Where hydrogen from electrolysis is used as a fuel or feedstock, about two-thirds of the input energy is available. When powering vehicles, the energy losses from primary energy to the wheel are greater with hydrogen than for battery-powered vehicles, but hydrogen vehicles do better in overall energy efficiency than conventional diesel or petrol vehicles.

The conversion losses mean that producing hydrogen from fossil fuels intrinsically makes most sense where a particular activity relies on hydrogen, such as for chemical processes, which to date are the largest global use of hydrogen; and for replacing diesel in heavy transport and machinery.

Another set of difficulties lies in storing and transporting hydrogen. Transporting hydrogen in large volumes requires combinations of extremely high pressures and extremely low temperatures, or immersion in chemical mediums. Transporting hydrogen in pipelines requires special materials.

The cost of shipping hydrogen from Australia to Europe is expected to be over US$2 per kilogram of hydrogen, in the absence of further technological advances. This is similar to expected future production costs of hydrogen, so shipped hydrogen is at a large cost disadvantage compared to locally produced hydrogen.[3] Storing hydrogen in chemicals for transport is cheaper, but the required chemicals are toxic and pose a risk in transport. Conversion of hydrogen to ammonia allows much cheaper transport, but the uses of ammonia are more limited than hydrogen, unless it is reconverted, which once again adds cost.

The potential game changer is the push for net-zero emissions, which will require decarbonisation of global energy and industrial systems. If most energy uses need to be zero-emissions or very low in emissions, large opportunities emerge for hydrogen. This is what will drive the likely push to hydrogen.

In a world that moves to net-zero emissions, a premium will be paid for clean fuels, and industrial processes will tend to use cleanly produced hydrogen rather than gas or coal. Falling costs of renewable energy generation and hydrogen production technologies mean that the cost of green hydrogen will fall.

Estimating the size of the future global market for green hydrogen at this point is guesswork, because it is unclear to what extent hydrogen will displace fossil fuels in different applications, or to what extent direct electrification will obviate the need for hydrogen and hydrogen-based fuels. The International Energy Agency's scenario for a transition to net-zero emissions has global low-carbon hydrogen production at 17 EJ (exajoules) in 2030 and 60 EJ in 2050, assumed to be two-thirds green and one-third blue. In comparison, total final energy consumption is around 420 EJ per year today, and in the IEA net-zero scenario it is assumed to gradually fall in future to 390 EJ in 2030 and 330 EJ in 2050.

Scenarios that do not assume that the world achieves net-zero emissions in the energy sector by mid-century have much less hydrogen production. For example, IEA modelling of currently announced climate pledges has future hydrogen use at less than half compared to the net-zero scenario.

Such projections need to be taken with a large grain of salt as they depend on assumptions about the future costs of different energy sources, the costs of producing and shipping hydrogen, technologies for fuel substitution, government policies and more.

Grey, blue or green: How clean is hydrogen?

Different production methods for hydrogen have vastly different emissions footprints. It can be made as a clean fuel, by producing it through electrolysis, splitting water by means of an electrical flow. If the electricity used is generated from renewable energy sources such as wind and solar power – or nuclear power – then there are no greenhouse gas emissions involved in its production and use.[4] Hydrogen made through electrolysis using renewable energy is termed 'green' hydrogen. Green hydrogen is, in effect, carbon-free electricity stored in molecular form.

Hydrogen can also be produced from fossil fuels. It can be made from gas through a process called steam methane reforming, or from coal. The production processes releases carbon dioxide, and if this is released into the atmosphere, the benefits of zero emissions at the point of use are negated. Producing hydrogen from gas without capturing the carbon dioxide results in what is called 'grey' hydrogen – and from coal, 'black' or 'brown' hydrogen.

In 2020, over 99 per cent of the world's hydrogen was produced from gas or coal.[5] It has been cheaper to use these processes than to produce hydrogen through electrolysis.

But hydrogen from fossil fuels is among the most emissions-intensive of all fuels. This is because the energy losses in the conversion mean that more fossil fuel is used than from burning gas or coal directly. Grey or black/brown hydrogen is typically 30 to 60 per cent higher in greenhouse gas emissions intensity, measured as the amount of greenhouse gas emissions per unit of energy, than direct combustion of gas or coal. The carbon dioxide emissions intensity of grey, black and brown hydrogen are around 74, 157 and 170 $kgCO_2$ per EJ of energy respectively, or around 9, 19 and 20 $kgCO_2$ per kg of hydrogen.[6] In addition, oftentimes sizeable emissions of methane are produced during the extraction, processing and transport of gas and coal.

Therefore, conventional hydrogen production chains are out of the question in a world that cuts emissions to address climate change.

The emissions intensity of fossil-fuel-based hydrogen can be lowered by capturing the carbon dioxide released in the processing of gas or coal to hydrogen and pumping it to underground storage reservoirs. The resulting product, hydrogen from fossil fuels with carbon capture and storage (CCS), is termed 'blue hydrogen'.

In Australia, blue hydrogen is often called 'clean' hydrogen, with little regard to the actual emissions intensity of the production system.[7] Blue hydrogen cannot be zero emissions. It is technically impossible to capture all the carbon dioxide in the production process, and processes geared to capture 90 per cent or more do not always perform to this standard in practice. Methane emissions during the fossil-fuel extraction and processing stage are not addressed through CCS.

Even with very high rates of carbon capture and storage, the remaining emissions in the production chains of coal- and gas-based hydrogen are large. At carbon dioxide capture rates of 90 per cent, the hydrogen produced may have a greenhouse gas emissions intensity of between a quarter and half, and sometimes more, of the underlying fuel. In part this is because of methane emissions in the fossil-fuel supply chain.[8]

Consequently, blue hydrogen is not the long-term answer for a decarbonised world energy and industrial system.

Blue hydrogen is often seen as a necessary step in the evolution of a global hydrogen supply chain, as today it is typically cheaper to produce than green hydrogen, assuming fossil fuel prices at longer-term averages. Many countries' hydrogen strategies assume that the hydrogen systems of the future will start with blue or even grey, black or brown hydrogen, and move to zero-carbon hydrogen gradually over time. In 2021, an analysis of twenty-eight national hydrogen strategies found that twenty-four of these followed a 'scale first, clean later' approach.[9]

It is easy to see why many consumers of hydrogen do not place appropriate importance on the greenhouse gas intensity of hydrogen production. The emissions arise and are accounted for at the production level, while for the consumer it is a perfectly clean fuel with zero accounted emissions (aside from hydrogen leakage, which is not currently captured in emissions accounting), no matter how dirty the production process. Where hydrogen is traded between countries, the emissions are reflected only in the exporting country.

For Australia, as a potential large-scale exporter of hydrogen, establishing a hydrogen industry that uses gas or coal as the feedstock would result in sizeable additional emissions. It would make it harder to achieve any national emissions reductions target.

Australia's 2019 National Hydrogen Strategy is premised on blue hydrogen providing the bulk of Australia's future hydrogen production. In fact, the idea of turning gas into hydrogen was wrapped up with the

Morrison government's political rhetoric of a 'gas-led recovery' from the economic downturn resulting from the COVID-19 pandemic.

A blue hydrogen industry would lumber Australia with a potentially large additional source of emissions. And once established, such an industry would be likely to persist, as its capital costs would be sunk and it might be able to exert pressure on governments, for example for exemptions from future penalties on emissions. If they were closed before the end of their technical lifetime, blue hydrogen production facilities would become stranded assets or partially wasted investments.

The answer, therefore, is green hydrogen. The future of hydrogen is for electrolysis driven by renewable energy, typically wind and solar power. This reality is dawning on many corporations and governments.

Figure 5.1 The emissions intensity of different fuels and hydrogen from different production methods.

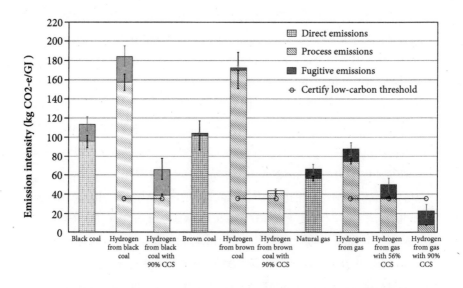

Source: T. Longden, F.J. Beck, F. Jotzo, R. Andrews & M. Prasad, '"Clean" hydrogen? Comparing the emissions and costs of fossil fuel versus renewable electricity-based hydrogen', *Applied Energy*, 306, 2022, pp. 118–45.

The European Union, currently considered the largest source of hydrogen import demand in the medium term, is planning to put in place a regulation that would set a ceiling for the emissions intensity of imported hydrogen, which in practice would allow only green hydrogen. The EU 'REPowerEU' initiative explicitly supports 'renewable' hydrogen, with a target import of 10 million tonnes of green hydrogen by 2030, alongside the same quantity produced in the EU. Other hydrogen-importing countries could follow suit.

Hydrogen production costs

Estimates of the total production costs for blue hydrogen from gas in large-scale production systems range between US$1 and US$3 per kilogram of hydrogen for gas-based hydrogen, with a median from a sample of studies of just over US$2 per kg.[10] Production costs are somewhat higher for coal-based hydrogen. They are obviously higher if a carbon penalty is imposed on residual emissions, and carbon prices even at low levels tip the cost balance in favour of blue over grey, black or brown hydrogen.

The spread in cost estimates is to a large extent due to different assumptions about fuel prices. At the time of writing, gas and coal prices are far above historical averages, which would result in much higher production costs for fossil-fuel-based hydrogen. For example, an increase of $10 per GJ in the gas price would add around $1.70 per kg to the cost of producing blue hydrogen with 90 per cent CCS. Wholesale gas prices in southeastern Australia were on average around $6 per GJ over the decade from 2011 to 2021; spot prices increased dramatically in the first half of 2022.

Green hydrogen is estimated to have median production costs in the order of US$3 to US$4 per kg under present-day conditions, with many studies arriving at higher or lower costs. It is lower in locations with access to low-cost renewable energy.

A higher carbon price will benefit green hydrogen in the cost comparison, but the larger effect will be from reductions over time in the cost of electricity from renewable energy, and the cost of electrolysers. These are the main cost factors for green hydrogen, and both are on a declining long-term trajectory.

The cost of both solar panels and wind turbines has fallen dramatically with the scaling-up of global production volumes and technical improvements. The supply chain pressures that the world is experiencing in 2022 are putting upward pressure on equipment and construction costs, but continuation of cost reductions is likely in the medium to long term. The scale effect of very large renewable energy installations will further drive down costs, for example through novel technologies for automated installation of solar panels.

Figure 5.2 The production cost of different fuels and hydrogen from different production methods.

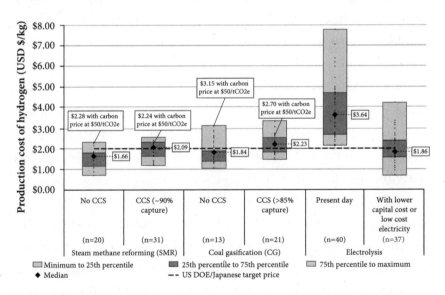

Source: T. Longden, F.J. Beck, F. Jotzo, R. Andrews & M. Prasad, '"Clean" hydrogen? Comparing the emissions and costs of fossil fuel versus renewable electricity-based hydrogen', *Applied Energy*, 306, 2022, pp. 118–45.

Electrolysers are set for a major drop in prices. To date, they are used in niche applications and produced in small quantities. Mass production will allow much lower production costs, through larger scale of units, process improvements and greater competition between producers.

Taking these factors into account, the costs of producing green hydrogen are likely to rapidly fall below those of blue hydrogen in locations where renewable energy can be produced at low cost.

Where will green hydrogen be made?

The physical prerequisites for cost-effective large-scale hydrogen production are straightforward: a large-scale supply of low-cost renewable energy, which typically means a great deal of sunshine and good wind speeds (preferably in combination); available land to build large arrays of electrolysers and the energy supply to run them; and access to industrial ports or a geography that allows the building of suitable ports.

Water supply is also needed, but the required quantities make its contribution to costs small relative to energy, making it for example feasible to desalinate seawater if fresh water is unavailable.

There can be variations on the energy supply. It could be hydropower, offshore wind or floating solar panels. However, the lowest-cost energy supply at very large scale will usually be from solar and wind on land.

Large areas of Australia perfectly meet these conditions, including the Pilbara region and coastal areas in southwestern Australia, central Queensland and South Australia.

Australia is of course not alone in boasting such potential. Similar geographies are found in northern and southwest Africa, Chile and the Middle East. Many of these locations are also closer to European markets, where much of the early demand for green hydrogen is likely to be.

Still, Australia is in pole position to become a location for green hydrogen production for export, because it marries the physical comparative advantage with institutional and economic advantages.

Australia has a cost advantage over developing countries in capital-intensive industries because the required interest rate (the cost of capital) is substantially lower in Australia. Investments in Australia carry lower risks because of its relatively stable political, investment and trade framework.

Australia also has a track record of successfully building large resource projects, including in the energy sector. The closest comparison with future hydrogen industries are the liquefied natural gas processing and export terminals. This gives Australia the required local expertise in engineering, project implementation, financing, regulatory approval, and physical and human capital infrastructure. These factors provide investors with confidence, which is especially important in a new industry where there are unavoidable project risks.

These risks are greatly amplified in countries where state institutions are weaker and where the industrial-technological system is not as deep. This is the case for most of Australia's main potential competitors. The risks are amplified in cases where major new infrastructure including ports needs to be built. Such infrastructure typically has long lead times, with risks of delays or non-completion. Political instability deters investors in some of the locations that may compete with Australia.

In a world of fracturing geopolitics, large-scale future trade relations will once again depend to a greater extent on political relations and shared norms, traditions and belief systems. For potential hydrogen-importing countries in Europe, Australia is an attractive option in this light also.

These factors beyond production cost matter more for hydrogen than for many other commodities because of a somewhat symbiotic relationship between producers (exporters) and consumers (importers), at least in the early stages of a hydrogen system. Building up a hydrogen use system, for example in Europe's chemical industry, demands reliable supplies. Importers will want to lock in supply contracts with reliable partners. Likewise, to commit to investment, producers and their suppliers of finance need confidence that purchase contracts will be honoured.

These factors suggest that hydrogen trade will likely be built up based on long-term contractual arrangements between trusted partners. They may see active involvement of the importing country in setting up supply chains, for example in the form of companies that are technology leaders in electrolysis-delivering equipment. They may involve foreign direct investment in, or joint ownership of, production facilities. They may even involve direct government involvement of various forms.

Australia is therefore well positioned. The most likely early partners at scale would be European countries including Germany: they are under pressure to diversify from Russian gas; they are strongly committed to decarbonisation, which translates to a strong preference for green hydrogen; and there is limited opportunity for cost-effective hydrogen production in Europe.

Likely features of large-scale green hydrogen projects

The specific form that large green hydrogen production systems take will vary with local conditions, but there are some shared features that are likely to prevail across Australia.

Typical large-scale green hydrogen production systems on mainland Australia would contain co-located solar photovoltaic and wind power, powering electrolysers located close to an industrial port. The hydrogen is prepared at the port, through liquefaction under pressure, immersion in a chemical carrier agent, or conversion into ammonia. The electricity generation may be adjacent to the coastal processing plant, or some distance away (as in central and north Queensland locations). In Tasmania, green hydrogen systems might be powered exclusively by wind backed by hydroelectric power.

Electricity supply for electrolysis may be standalone. It could be augmented by energy storage, for example through pumped-hydro power, where this is justified by cost saving from more intense utilisation of the electrolyser capital.

Hydrogen supply systems could also be connected to the electricity grid, saving costs in both. The two-way integration can increase utilisation rates for the electrolysers and reduce the need for energy storage and peaking capacity on the grid, because the grid can draw on the power supply built for the hydrogen system.

Grid connection of hydrogen production systems raises some complex questions for the sharing of costs and benefits, and for pricing. The transmission line and connection equipment can be a major cost factor. Whether and in what proportion the cost is borne by the hydrogen development or the grid should depend on expected benefits to each, rather than on regulatory rules established for different circumstances.

Grid connections of electrolysers present challenges for certification as 'green' hydrogen, which is essential for exports to high-value markets. Accounting and certification systems need to attain high levels of sophistication. There is a risk that overly strict or simplistic requirements on hydrogen imports, as are for example under development by the European Commission, could preclude efficiencies from grid connection of hydrogen production systems.

Hydrogen production systems require immense amounts of land for solar panels and wind turbines, and large areas for electrolyser banks and hydrogen processing and storage facilities near the coast.

Facilities will typically be located in areas where the opportunity cost of land is low. Electricity generation and electrolysis do not produce pollution, and water supply will usually be adequate.

But projects of this scale will generally have adverse effects on local environments, including biodiversity. These effects come about from construction, including the building of access roads and changes to local conditions for fauna and flora that arise, for example, from shading from solar panels over very large areas. The keys are to choose sites not only for minimisation of production costs but also minimisation of environmental footprint, and to design installations to take account of local environmental conditions.

In addition, some forms of processing for shipping of hydrogen and derivative products can involve local pollution or the risk of accidents with toxic materials.

Much of the land required will be subject to differing strengths of First Nations rights and interests, including through native title and land rights legislation. Benefits for First Nations will be more likely if Traditional Owner groups are engaged early and are well-resourced and well-informed so they can meaningfully engage in development including through agreement-making processes.

Economic benefits need to be fairly shared with local communities, including and especially First Nations people.

Using hydrogen in Australia: Another export story

Hydrogen is also likely to be used in Australia. Scale depends on whether an export industry for using hydrogen for energy-intensive processing of commodities emerges. The main candidates are iron metal and steel (see Chapter 6), synthetic fuels and fertiliser production.

Hydrogen is likely to be used in some transport applications in Australia. Hydrogen is likely also to be used to a small extent in the power system, as an energy storage medium in specific cases where this is cheaper than energy storage in batteries or pumped-hydro facilities. These uses are small compared to the potential for export industries.

Hydrogen can also be mixed in with the general gas supply. This is limited to a small share for technical reasons and would constitute a low-value use of hydrogen. For energy use by households and general businesses, decarbonisation is better achieved through direct electrification.

Decarbonisation of transport can be achieved through electrification through batteries, electrified rail and zero-emissions onboard fuels. Battery electric vehicle technology is set to win a lion's share of the market for clean on-road transport. It is shaping up as the technology of choice for cars and for at least part of the truck and bus fleet, owing to the rapidly falling cost of batteries, higher overall energy conversion efficiency and lower capital requirements for refuelling. Hydrogen may become the fuel of choice for long-distance and point-to-point heavy goods transport, including in heavy trucks and on rail lines that are not electrified. Hydrogen-powered flights may also become possible. It may also become competitive for transport and some stationary energy uses such as mining in remote locations, where onsite renewable energy generation is preferable to shipping fuels over long distances. Whether hydrogen or batteries are the cheaper storage option depends on the specific circumstances.

In shipping, ammonia (which can be produced from green hydrogen) is likely to be part of decarbonisation. Ammonia can be burned in ship engines. Australia can supply a large international cargo shipping market.

Hydrogen could also be used as an intermediate step in the production of synthetic liquid fuels to replace diesel and aviation kerosene. Synthetic zero-emissions fuels, or e-fuels, using electricity as the energy input could become a very large industry for Australia; for domestic and international transport with fuelling in Australia; and as an export commodity.

Ammonia is the critical input in the production of nitrogenous fertilisers and some other chemical processes. It can be produced from green hydrogen with zero emissions, rather than natural gas with high emissions. One of the world's first renewable ammonia production plants is in preparation in the Pilbara. The value of the global ammonia market is in the order of $100 billion per year. That will grow massively with progress towards the zero-emissions global economy. Australia could achieve a sizeable share of the global market.

The largest prize for hydrogen is likely in iron and steel production ('green steel'). Primary steel is made by combining iron ore and additives with coking coal in blast furnaces. Steel production is responsible for 7 to 9 per cent of global greenhouse gas emissions, and this will need to be removed as the world pushes to net zero emissions. Part of the answer is to reduce steel use. Part is to increase the share of steel recycling, which uses electricity, not coal. But the world will continue to need large amounts of primary steel, made from iron ore. This requires new processes, and currently the most economical ones use a combination of hydrogen and electricity to reduce the iron ore to iron and produce steel. These processes are in the early stages of commercialisation.

For Australia, a tantalising prospect is that part of the expansion of the world's future iron processing and steel production could be located in Australia. Australia is the world's largest producer of iron ore (37 per cent of global production) and the largest exporter of metallurgical coal. It produces only a tiny amount of steel. Most iron ore and coking coal are exported, with most of the exported iron ore going to China.

Australia has obvious potential to be a location for a future green iron industry, because it can produce practically unlimited amounts of hydrogen at comparatively low cost. The same goes for renewable electricity for steelmaking. Iron-ore mining is concentrated in areas (including the Pilbara in Western Australia) which are also highly suitable for large-scale green hydrogen production.

Some arithmetic illustrates possible magnitudes:[11] if all iron ore exported from Australia (in 2019) were processed in Australia, this would produce around 540 million tonnes of steel per year, requiring 2100 TWh of electricity per year to make 33 MT of hydrogen per year. This is more than seven times the current total annual electricity generation in Australia. About 1000 GW capacity of wind and solar power would be required. The investments would be in the order of trillions of dollars. Most value would be added in Australia. The value of Australia's iron-ore exports is in the order of $100 to $150 billion per year depending on prices; the value of the steel produced from it is several times larger, if exported in its simplest form, crude steel. The value added from processing would dwarf the value of current coking coal exports. One option for this might be to produce an intermediate product, such as pig iron, in Australia, for export and processing into steel in traditional steel-producing locations in the industrial heartlands of East Asia, Europe and America.

Processing a quarter or half of Australian exports of iron ore into metal would build an immense industry.

The world has ample capacity in traditional blast furnace steel plants, and a green steel industry requiring large upfront investment will compete with existing plants for which only running costs is a factor. This will tend to slow the transition to green steel. However, it is likely that demand for green steel will arise and drive the establishment of a green steel supply industry. Some car manufacturers are already planning to produce some vehicle types using green steel, as a 'greener' product. In other industries, green steel might become the choice as a result of policies, for example regulation of the maximum amount of embedded emissions in the construction of buildings.

The role of governments

What can and should governments do to foster the emergence of a hydrogen industry?

The single most important aspect is policies to penalise greenhouse gas emissions and incentivise the uptake of zero-emissions alternatives, in particular carbon pricing as well as emissions standards for fuels.

For potential hydrogen-exporting countries such as Australia, the policies of energy-importing countries matter greatly. The European

Union's emissions trading system now imposes a sizeable price on greenhouse gas emissions, providing incentives for EU industry to switch from fossil fuels to renewable (as well as nuclear) energy and industrial inputs.

The decisive shift in demand is from gas to hydrogen, including to supply energy for European industry. Russia's invasion of Ukraine and the resulting move away from pipeline gas imported from Russia to Europe accelerates the demand for hydrogen in Europe, particularly in Germany.

European countries are preparing to subsidise initial deliveries of green hydrogen through competitive tender and contract-for-difference arrangements.

In Australia, a price penalty on carbon emissions would likewise be useful, to guide the emergence of an efficient and sustainable hydrogen industry. A price on emissions would make blue – and to an even greater extent grey, black and brown hydrogen – less competitive relative to green, by reflecting the costs their emissions impose on the community.

Governments at both federal and state levels have important roles in investment facilitation. They can make sure that regulatory requirements can be fulfilled without undue delays. They can help facilitate industry growth through spatial and infrastructure planning, and provide a general welcoming stance to the green hydrogen industry. Governments also have roles in creating and implementing certification schemes, allowing producers to demonstrate that their hydrogen has zero or very low emissions. And they need to fund applied research on technological as well as economic, social and regulatory aspects of a future hydrogen industry.

The federal government has a critical role in establishing international market access for zero-emissions products from Australia. Both federal and state governments can provide fiscal support for pilot and first-of-a-kind projects, recognising the risks taken by technological pioneers that confer benefits on the rest of the community.

Policy interventions that involve discretionary decisions by officials favouring particular projects, for example through subsidies or tax exemptions, should be avoided. These are especially costly for the Australian community when provided by state governments in competition with other state governments. Efficient allocation of investment requires a level playing field across the country. Necessary recognition of the costs of emissions and innovation in low-emissions technologies should be provided by mechanisms of general application.

The federal government, in particular, has an important influence on how attractive Australia is as a hydrogen-trading partner, through its commitments and actions on climate change. Governments in energy-importing countries that are committed to climate action will prefer to support hydrogen investments in countries that are like-minded on climate policy and act accordingly. The adoption of an internationally acceptable national emissions target and continuous strengthening of both the target and efforts to achieve it are key.

What is the national benefit from a hydrogen economy?

Where the possibility exists for the creation of potentially very large-scale energy and energy-intensive export industries, with extremely large revenues, it may seem that it is self-evidently desirable to make them happen.

But we should ask how the national interest is best served. Will Australia be better off as a result of a hydrogen industry? Who will reap the benefits, who will wear the costs?

Past experience with resource industries in Australia is that they have provided some national economic benefits, with large gains to many; but that they overall have had environmental and sometimes also social costs, and that the benefits have typically been unequally distributed in favour of the owners of capital rather than the community at large.

The birth of a new industry offers the chance to do better.

First, let's consider the potential financial benefits. Large-scale resource developments in Australia are almost always owned by private industry, and effective tax rates tend to be relatively low. As a result, a large share of profits typically goes to shareholders, including those overseas, while the public receive a relatively small share of the pie via their governments. The liquefied natural gas (LNG) industry is a case in point. Attempts at higher taxation, such as a super-profits tax on resource industries, have often failed due to the outsized political influence of vested interests.

Lessons from resource industry taxation apply in some ways but not in others. A green hydrogen draws on inexhaustible renewable energy, not finite physical resources. Similarities are that large projects of this kind have potential for damage to local environmental and Indigenous values. Their scale and associated opportunities for oligopolistic control can result in excess profits and risk capture of policy and regulatory processes. Appropriate analysis and policy development is needed from the start.

Next, we should look at the social and cultural benefits and costs, including regional development. Hydrogen production systems will bring very large investments to regional and remote areas. This is on the whole desirable, especially since the coal industry and then the gas industry will be phasing down in rural and regional Australia. Hydrogen production will bring obvious regional economic benefits; however, the number of jobs and thus the direct benefits to local communities are typically quite limited in developments of this kind, as the degree of mechanisation and automation is very high.

Questions around how such facilities would affect Indigenous communities, socioeconomically, environmentally and culturally, are critical considerations. First Nations people and communities must be involved in the decision-making, including about siting, design and implementation, to ensure that projects proceed in ways that best reflect their priorities. Where the direct impacts are negative, ways need to be found to arrange project siting and implementation to minimise adverse effects and to maximise the benefits to First Nations communities.

In this light, benefit sharing for new hydrogen industries is of paramount importance. The national interest would be best served by securing a significant share of the value and profits created for the public purse, to allow for investments that are in the community interest. Appropriate and ambitious regulatory requirements also need to be placed on the developments.

Finally, let us consider the environmental impact. Creating a large-scale hydrogen industry in Australia will be damaging to some extent for parts of Australia's environment. The standard computation is to weigh up such local environmental impacts against the financial benefits. But there is another, more fundamental consideration. That is the environmental effect on the world overall. The question then is: would building the hydrogen industries that are needed for decarbonising the world economy, in another part of the world, impose greater or lesser local environmental damages than building them in Australia? The answer may well be that Australia is a good location from a global point of view.

By establishing a green hydrogen industry, Australia will help the world to displace fossil fuels at large scale, and consequently to reduce emissions and limit climate change. This will be the direct opposite of Australia's historical contribution as a provider of fossil fuels. Hydrogen is a way to use more of Australia's near unlimited potential for clean energy – and make it available to the world.

6

DECARBONISING CHINA'S STEEL INDUSTRY

Ligang Song

In September 2020, Chinese president Xi Jinping announced China's commitment to peak carbon emissions by 2030 and to achieving carbon neutrality by 2060 (the 'dual carbon strategy'), as a contribution to the global defence against climate change. The decarbonisation of China's steel industry is essential to reaching an early carbon peak over the next few years and neutrality in the coming decades.

These goals require an unprecedented rate of decarbonisation – a huge challenge. Developed countries in which emissions peaked in the 1990s (nineteen countries) and developing countries where emissions peaked in the 2000s (thirty-three countries), had a fifty-to-sixty-year transition to carbon neutrality. China seeks to achieve this same goal within the next forty years.

The reduction of emissions from Chinese steel production is globally significant. China accounted for 53 per cent of global steel production in 2019 and was estimated to contribute more than 60 per cent of global steel emissions. The rapid growth in total steel output and its high carbon intensity together made China's steel industry a significant producer of carbon emissions. By 2019, steel carbon dioxide emissions totalled 1574 million tonnes. Steel accounted for 17.5 per cent of the country's total emissions, second only to power generation. Chinese iron and steel production contributed over 4 per cent of total global emissions from all sources in all countries, or more than three times Australia's current total emissions. Australia supplies about three-fifths of the ore used in primary steelmaking in China. Using Australian renewable electricity and hydrogen produced from renewables to convert that iron oxide into iron metal and steel would reduce global emissions by around 2 per cent – almost twice as much as Australia eliminating its own emissions. That would be a major step towards achieving Chinese and global carbon goals.

There is great potential for China's steel industry to reduce its carbon emissions, but the pathways to decarbonisation have not yet been defined and are the subject of much debate. The carbon emissions intensity of the steel industry in China was 1.8t CO_2/tonne in 2019, in contrast to 1.3t CO_2/tonne in the European Union and 1.5t CO_2/tonne in Japan and South Korea. The difference stems mainly from China making a much smaller proportion of its steel from scrap than established industrial countries that have a longer legacy of large-scale use of steel. BHP says that 'passive abatement' – using a growing scrap supply and taking advantage of the shift to zero-emissions sources in China's electricity system – could reduce emissions in China's steel production between 2019 and 2050 by 40 per cent.

In addition, a 2021 BHP study showed how China's steel industry could cut emissions by another 20 per cent by 2050 (compared to 2019) through proactive changes, including using higher-grade ores and coking coals and hydrogen steelmaking; technological upgrades and the use of renewable energy in a wider range of processes; and carbon capture and storage.[1] Yet such significant action would not be enough for the steel industry to meet the country's carbon-neutral goals – and in any case cannot be taken for granted, since the detailed application of new approaches has not yet been carried out and timetables for their implementation are not resolved. Although each individual measure plays a part in reducing total emissions or its intensities, their interaction is complex and affects total emissions reductions.

This chapter provides a brief overview of the development of China's steel industry and discusses the likely trajectory of its future growth. It highlights challenges in decarbonising the industry. It explores implications of various options for Sino-Australian trade in steelmaking raw materials – including how imports of iron metal made from renewables-based hydrogen can reduce the costs of decarbonising steel production. Finally, it discusses the wider implications of decarbonising steel production for sustainable development in China.

Unprecedented development of China's iron and steel industry

China's iron and steel industry has been regarded by the Chinese government as a pillar of development since the beginning of central planning in the 1950s. Over the past few decades, iron and steel production has grown at a pace and scale that has no near precedent in global economic history. After gradual reform late last century and then dramatic expansion driven mainly by rapid growth in demand from industrialisation and urbanisation, the industry has come to produce more steel and to use and import more iron ore than the rest of the world together. Total output has stabilised at around 1 billion tonnes per annum in recent years. This was predicted by Australian economists more than a decade ago. Figure 6.1 shows how China's per-capita steel production has risen since the 1950s.

China's steel industry expansion has been exceptional by global standards (Figure 6.2). After reaching peak levels of production, the United States, Japan and Germany maintained relatively high levels of

Figure 6.1 China's per-capita crude steel production, 1957–2020.

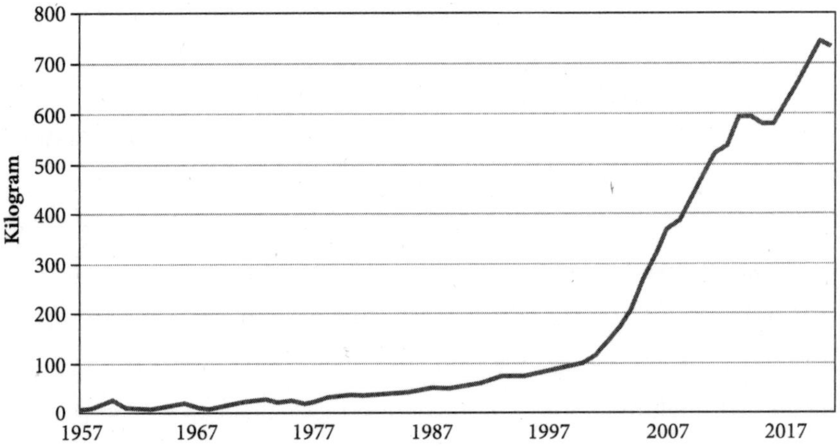

Source: Author's construction from CEIC Database.

steel production over time, while France and England declined steadily. Steel production fell substantially in Russia when the former Soviet Union collapsed in the early 1990s but showed an increase once the economy returned to growth from a lower base. The steel industry in India began to take off at the end of the 1990s, with total steel output being comparable with other major steel producers but small compared with China.

In 2020, China's steel industry produced over 1065 million tonnes of crude steel, accounting for 56.4 per cent of the world's total. 995 million tonnes were consumed domestically, making up 56.2 per cent of total global steel consumption. The construction sector accounted for most of this (58.3 per cent), followed by machinery and equipment production (16.4 per cent) and automobile manufacturing (5.4 per cent).

China now relies heavily on imported iron ore and coking coal (Figure 6.3). In 2021, China imported 1.1 billion tonnes of iron ore with an average price of US$164 per ton. Imports made up 60 per cent of China's total iron ore consumption. China accounted for more than

Figure 6.2 Total steel output in selected economies, 1870–2021.

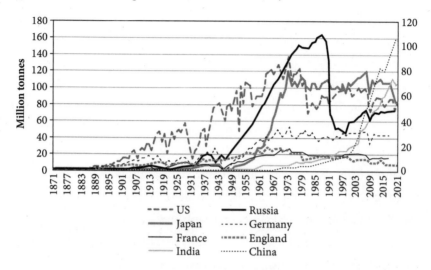

Source: For 1871–2015, Baidu Wenku, The crude steel production volume of world's major economies, 1871–2015, 2016; For China 2015–2021, CNKI, https://data-cnki-net-s--ncu.cxkjj.top/YearData/Analysis; For other economies 2015–2021, Statista, www.statista.com.

Figure 6.3 China's imported iron ore (million tonnes), 1985–2017.

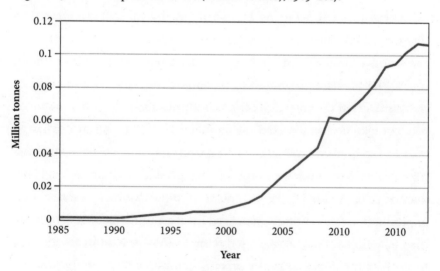

Source: CNKI, https://data-cnki-net-s--ncu.cxkjj.top/YearData/Analysis.

80 per cent of Australia's iron ore exports and a quarter of its coking coal exports in 2020. From 2000 to 2017, China's compound growth rate of iron ore imports was over 16 per cent. Imports for the rest of the world fell. China accounted for over 60 per cent of global iron ore imports from 2010, up from less than 20 per cent before 2002.

Three main factors contributed to extraordinarily rapid growth of steel production. First, rapid urbanisation and incomes growth increased demand. Between the late 1970s and 2021, the proportion of the population that lived in urban areas rose from about 20 per cent to about 64 per cent, and per-capita income rose from a few hundred US dollars to more than US$12,000 per annum. Industrial growth shifted from concentration on labour-intensive light industries to capital-intensive manufacturing, and then with an increasing component of high-tech industries. Demand for steel grew most strongly in the capital-intensive stage, with its focus on infrastructure and heavy industry. The rapid expansion of the steel industry laid the foundation for China becoming the world's largest producer of manufactured goods, and largest trading nation. China built a modern urban and transport infrastructure, contributing a majority of the world's investment in highways, high-speed trains and modern airports over this period.

Second, ongoing market-oriented reforms and investment in upgrading plants contributed to technological progress and productivity growth in the steel industry. Labour productivity and total factor productivity (TFP) increased substantially. Between 1998 and 2020, the average integration ratio (moving hot iron continuously from the blast furnace into steelmaking and rolling final products) for crude steel production of the industry – one of the most important indicators for level of production technology – increased from around 40 per cent to more than 99 per cent. Market-oriented reform of the steel industry also led to a substantial increase in the private sector's involvement in producing steel products.

Third, the integration of the Chinese economy into the global economy, especially since China's accession to the World Trade Organization in 2001, linked China into global production networks (GPNs). Increasingly, industrial materials such as iron ore, coal and other mineral commodities were drawn from international sources, supporting exports of industrial goods with high metal intensities, such as machinery and transport equipment. According to WITS, in 2001 China's exports of machine and electrical goods were US$85 billion, accounting for 31.9 per cent of total exports.[2] By 2019, exports had increased to US$1087 billion, representing 43.5 per cent of exports (Figures 6.4 and 6.5).

The prospects for China's future steel demand

China's per-capita income reached US$12,000 in 2021 – the level that typically marks the end of heavy industries playing an expanding role in the national economy. The manufacturing share in total GDP stood at about 28 per cent in 2021, and this is likely to be maintained for some

Figure 6.4 China's exports of electrical, electronic equipment, 1992–2020.

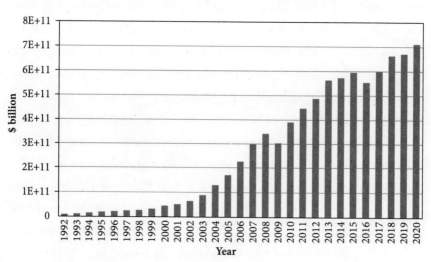

Source: Trading Economics, https://tradingeconomics.com.

Figure 6.5 China's exports of machinery, including power generation, 1992–2020.

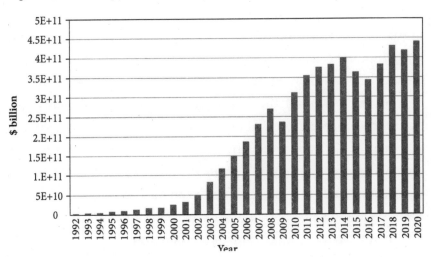

Source: Trading Economics, https://tradingeconomics.com.

time before it falls as the economy shifts into a bigger role for services as China's per-capita incomes continue to increase.

China's urban population is now 64 per cent of the total. The government aims to lift this to 70 per cent or higher in the next phase of growth. Continued urbanisation means further expansion of two metal-intensive sectors: housing and infrastructure. Both sectors have been undergoing structural adjustments in recent years. This has gone furthest in housing, where the easing of housing prices led to the liquidation or financial restructuring of some real estate companies. China continues to rely heavily on investments in such metals-intensive infrastructure as high-speed trains and air transport to maintain strong economic growth.

By the end of China's fourteenth Five-Year Plan, 2021–25, more than 95 per cent of cities with a population over 500,000 will be connected to the high-speed railway network, with trains running at 250 kilometres per hour and above. The length of high-speed rail will increase from 38,000 kilometres to 50,000 kilometres. With the construction of more than thirty airports nationwide, air transport capacity will rise from 1.4 billion trips per annum in 2020 to 2 billion in 2025.

China's exports will continue to shift to greater proportions of more capital-intensive and technologically sophisticated metal-intensive products, such as automobiles, machinery and transport equipment.

China's Belt and Road Initiative (BRI) has also played an important role in generating demand for steel, because investments in infrastructure is a core part of its programs. In 2020, more than half the investment in BRI countries was in manufacturing, construction, and electricity generation, with a total value of US$13.9 billion.

Increasingly, demanding domestic regulation is increasing steel production costs, reducing international competitiveness of Chinese steel and lowering its global market share. But exported steel only occupies 5 per cent of overall steel production, so this has a limited impact on China's steel demand. A reduction in steel exports helps China's environmental objectives, including emissions reduction. This is part of the reason why the Chinese government has sought to reduce steel exports by removing standard tax rebates on exports – analogous for GST rebates for exports in Australia – for steel producers.

In the longer term, when Chinese economy industrialisation has moved beyond its metal-intensive phase and urbanisation slows, steel demand growth will slow and then go into reverse. After China's urban population reaches 70 per cent, demand for new housing will decline and drag down demand for steel. Construction is currently the largest steel consumer. Construction demand is predicted to peak in 2024 and drop at a compound rate of 2.6 per cent annually from then. Machinery and equipment production is currently the second-largest steel consumption sector. It is expected to decline at an annual compound growth rate of 1.2 per cent as China moves towards a more service-driven economy.

With more advanced technology, machinery and equipment products are more durable, which means the demand for replacement drops steadily. The electric car's motor can be expected to travel many more kilometres, or several times more kilometres, without replacement. Electric cars contain less steel relative to value. The structure of steel demand also changes with technological improvement. When a higher fraction of production goes into manufacturing more technologically complex and efficient products, demand for steels with special fineness, strength and other high qualities increases. This lowers demand for steel quantity, but not necessarily total value.

McKay, Sheng and Song (2010) drew attention to a Kuznets curve for steel (KCS), with demand growing with income until a turning point is reached.[3] After that, demand falls with increased income. China's turning point was estimated at US$15,449 GDP per capita – a point China would reach during 2024 on its post-1980 trajectory of 7 per cent compound growth rate. At that time, China's steel demand per capita was expected to be over 700 and approaching 800 kilograms. That is higher than the peak in the United States, the Commonwealth of Independent States and Europe, but lower than in Japan, Korea and Taiwan. With the slowdown of economic growth in China in recent years, the turning point of China's KCS can now be expected much later than 2024.

Major factors shaping Chinese steel decarbonisation

Several policy approaches must contribute to reaching the goal of decarbonising the Chinese steel industry. Progress depends on how successfully China integrates these different policy approaches.

China has made substantial progress in structural reform, including in ownership (including private ownership) and corporate governance. The share of state-owned enterprises in crude steel production was 90 per cent in 2001, but less than 40 per cent in recent years.

Increased economies of scale may be conducive to reduction in emission intensities for the industry. The industrial concentration of the steel industry has risen (with the share of the ten largest firms rising from 35.9 per cent in 2016 to 41.5 per cent in 2021) but remains well below other major steel-producing countries, including Japan and Korea. World steel association figures measure the concentration ratio (CR3) – the share of the three largest steel companies – in the range of 50 to 90 per cent in the United States, Japan, India and South Korea. China's CR3 is only 17.3 per cent.

China encourages steel industry consolidation through mergers and acquisitions. This could increase industrial concentration. Chinese authorities have released a guideline encouraging domestic steel mills to consolidate to create world-class steel giants and accelerate production of high-quality products. Official guidelines suggest that the government will rely on the dominant enterprises in the industry to introduce pilot plants for production of stainless steel, special steel, seamless steel pipe and other special items. Financial institutions are urged to provide financial services to iron and steel enterprises pursuing mergers and acquisitions, structural adjustments, transformations and upgrading.

Supportive regulatory and policy environments are necessary to provide incentives for large emissions reductions: emissions trading schemes, pollution discharge standards, energy-saving mandates, and fiscal support for the introduction of new low-emissions technologies.

From the beginning of the fourteenth Five-Year Plan (2021–25), China has been promoting green development in the steel industry to realise carbon neutrality by 2060.

The Ministry of Industry and Information Technology's plan proposes that by 2025, China will have several immense steel conglomerates, and the top ten steel enterprises will account for 60 per cent of output. In August 2021, China's two biggest steelmakers, Ansteel and Ben Gang Group Corporation – agreed to merge. After the merger, Ansteel ranked third in the world, with an annual production capacity of 63 million tonnes of crude steel.

Changing the energy mix and raising energy efficiency is crucial to lowering steel industry emissions. In the next few decades, electric arc furnaces will become the largest source of steel supply. These convert scrap steel or iron metal ingots into crude steel, through application of huge amounts of electricity. In Chinese conditions, the electric arc furnace (EAF) allows the steel to be made with around 0.6 tonnes of carbon dioxide per tonne of steel, compared with several times that from use of iron ore in a blast furnace. About one and a half tonnes of carbon dioxide per tonne of steel come from reduction of iron using metallurgical coal in the blast furnace. There are additional emissions in making coke, converting molten iron to steel and rolling final products.

The 0.6 tonnes per tonne of steel falls as the supply of zero emissions power to the grid increases, potentially to zero. Chinese scrap supply will expand rapidly with the passage of time, as the stock of assets made from steel grows older. Decarbonisation of electricity supply allows zero emissions for steel made from scrap.

In 2020, EAF steel accounted for 10 per cent of China's crude steel production. This compares to 70 per cent in the United States and 30 per cent on average in the world. The government's plan aims to raise the proportion of EAF in total output to 15–20 per cent by 2025, and the use of scrap in all steel production to about 30 per cent. The scrap, which contains impurities, can be augmented in the EAF by pure iron metal, which increases total output and dilutes impurities. Iron metal produced from iron ore by direct reduction using hydrogen made from electrolysis using renewable energy can therefore expand the contribution of the electric arc furnaces while increasing steel quality. There are still emissions for as long as the electricity used in the electric arc is generated from coal, gas or petroleum. The large amounts of electricity required in the EAF increases demand for power and the challenge of decarbonising power supply, so there are advantages in importing some zero-emissions metal as steel ingots rather than iron metal. Imported green iron is likely to be much cheaper than making it in China with local renewable energy – just as it is cheaper for China to use imported rather than domestic iron ore.

At this stage, prior to development of large-scale international trade in iron ingots, the primary constraint on EAF steel production is the limited supply of steel scrap. In 2020, alongside a billion tonnes of steel production, only 260 million tonnes of steel scrap were available in China. By 2030, the annual supply of steel scrap is expected to reach 0.4 billion tonnes, sufficient to supply 25 per cent of predicted steel demand. In 2050, 78 per cent of China's energy consumption will be sourced from non-fossil energy and 60 per cent of steel supply from recycled EAF steel. This proportion could be increased almost without limit by importing iron in the form of iron produced by direct reduction using zero-emissions hydrogen.

Adopting carbon capture and storage (CCS) technology in the steelmaking industry may be able to contribute to emission control and reduction. According to Arasto et al. (2013), Rautaruukki's Raahe steel mill generated 4.5 Mt/Y in 2008, and CCS technology successfully captured 2.9 Mt/Y of carbon dioxide.[4] This is only partial capture. For application of CCS to be commercially attractive even when good sequestration sites are located conveniently to the blast furnace, an emissions price is in the range of 46 to 90 Euros per tonne of carbon dioxide emissions and electricity prices in the range 80 to 100 Eurodollar/MWh are required. In China, the emission cap-and-trade systems are at an early stage of development. The steelmaking industry is currently not covered, so there are no commercial reasons for using CCS in China's steel industry. However, with a higher carbon price and a lower CCS cost, the technology could make a contribution. In the meantime, with current carbon, electricity and other costs, expanding the EAF output through using local and imported scrap with imports of zero-emissions iron metal is a more likely route to large emissions reductions.

The overall increase in China's research and development (R&D) expenditure is leading to the use of new and more advanced technologies in China's steel industry. Official reports put the ratio of China's R&D expenditure to GDP at 2.44 per cent in 2021. In 2021, the global index of innovation published by the World Intellectual Property Organization ranked China's innovation capability twelfth among 132 countries.

China has made important progress in producing hydrogen, which could be used to replace metallurgical coal to achieve zero emissions in ironmaking. Traditionally, the steel industry uses coking coal to reduce the iron oxide (iron ore) to metal, with the carbon removing the oxygen from the ore to leave pure iron, with carbon dioxide as the waste produce. Using hydrogen as the reducing agent, the waste is water. Iron can be produced with zero emissions in this way.

The Chinese government strongly supports the use of hydrogen. In 2021, the fourteenth Five-Year Plan period encouraged innovation and technological progress on hydrogen steelmaking.

The primary method of green hydrogen production is water electrolysis. Bhaskar et al. (2022) estimate the levelised cost of hydrogen-based steel production in Norway at 40 per cent higher than the traditional blast furnace basic oxygen furnace (BF-BOF) method.[5] Since then, both electricity and metallurgical coal prices have increased by large amounts. Compared to the retail electricity price of US$61/MWh adopted in the analysis, China's average electricity retail price was US$96/MWh in 2019, indicating an even higher production cost.

Norwegian and Chinese electricity and coal costs have increased by large proportions since this analysis was completed. Production of green hydrogen in locations with favourable combinations of wind and solar resources can be undertaken at much lower cost. Australia – also the location of most Chinese iron ore imports – is the prime location from the point of view of physical resources. Electrolysis to produce hydrogen from water and electricity is readily curtailable, so intermittent renewable energy does not need to be backed completely by storage or thermal power. As a result, the cost of renewable energy into an electrolysis facility in favourable Australian locations with access to high-quality wind and solar energy resources, would be very much lower than those in the Norwegian study, and lower still than costs in China.

The high cost of international trade in hydrogen (with delivered cost in China more than twice as high as in Australia, compared to less than 10 per cent higher for metallurgical coal) suggests that the processing is undertaken more economically in Australia than in China using imported hydrogen.

Economy-wide hydrogen-based iron and steel production in China would require a comprehensive nationwide hydrogen transportation system. The hydrogen embrittlement effect means the hydrogen gas pipeline would need to be made from low-carbon materials, which more than doubles gas pipeline cost.

Despite the current high cost of hydrogen usage, the commercialisation of steelmaking through use of domestic green hydrogen could be achieved by a significant decrease in retail electricity price or technological progress. From 2008 to 2021, China's lowest solar grid purchase price dropped from 1.09 ¥/KWh to 0.1476 ¥/KWh and is expected to decrease further. While solar and wind energy are getting a larger share in China's electricity consumption, the retail electricity price will likely drop in the next thirty years. Meanwhile, the technology progress saved 22 per cent of storage tank production costs. The mixture of hydrogen and natural gas has been found to effectively avoid the hydrogen embrittlement effect that keeps it to transport through the natural gas pipeline. If any similar technological progress appeared in the production, compression, transport and storage process of hydrogen, zero-emission steelmaking would be achievable at the time.

Mike Sandiford (Chapter 3) suggests that there are economic advantages in producing the hydrogen in Australia to reduce Australian iron ore to iron metal. It is economically efficient for China to shift early-stage production offshore where low-cost renewable energy resources are close to iron ore mines. The dependence of Chinese steel industries on Australia – or, for that matter, of Australian exports on the Chinese market – would change in composition (from iron ore to metal) but not fundamentally. This would be conducive for Chinese and global emissions and for the health of the Australian and Chinese economies. It would require a return to productive and cooperative Sino-Australian relations.

Impact and significance of decarbonisation in the steel industry
Measures taken to decarbonise the steel industry and the likely outcomes will contribute in a significant way to China's 'dual carbon strategy'. The steel industry will greatly increase efficiency, reduce exports, remove emissions and move towards environmental sustainability. Use of recycled scrap will reduce reliance on new raw materials. The shared understanding of the necessity of low-carbon growth in the industry widens the scope for international cooperation.

Future demand for iron ore and coking coal will fall. The increasing stock of steel scrap and falling steel demand will substantially reduce use of iron ore. There will be a much larger fall in imports of coking coal: the EAF process doesn't require a reducing agent, and a substantial part of reduction will use hydrogen. After the commercialisation of hydrogen-based steelmaking, the demand of coking coal may be eliminated in the steelmaking industry. This change in demand could heavily affect the Australian economy, as coking coal and iron ore count for over one-third of total Australian exports.

Demand for electricity in China will rise. Demand for new EAF and hydrogen electrolysis will greatly increase – in China and abroad.

China's production capacity in electric and industrial equipment would allow it to make strong use of growing international demand.

Steel is an important upstream material for construction and equipment manufacturing industries, so steel prices influence their profitability. In the next decade, with the strengthening of the carbon cap-and-trade system, the cost of reducing emissions could push up steel prices and costs in downstream industries. However, in the longer term, the decreasing electricity retail prices, the broad application of EAFs and use of opportunities for importing green iron metal may drag down steel prices and benefit downstream industries. The associated increase in Chinese exports of renewable energy and hydrogen capital goods is likely to be immense.

A less export-oriented Chinese steel industry will significantly change the international distribution of steelmaking. In 2021, China exported 67 million tonnes of crude steel, 40 per cent less than 112 million tonnes in 2015. Since 2016, China has aimed to cut steel over-capacity by shutting down environmentally unfriendly or inefficient units. At the same time, the Chinese government began to restrain steel exports and encourage imports of steel products. The decrease in crude steel exports within the decarbonisation strategy will leave more space for other countries to develop their own steel industries. Developing countries, such as India and Thailand, could utilise this opportunity to increase their steel production. Australia has many economic opportunities, including beyond the ironmaking stage.

Given that it is the world's largest producer of steel, China's experience will have important implications for decarbonisation of the world steel industry. At the same time, China will continue to learn from those countries whose steel industries have made some breakthroughs in applying the more advanced technologies and adopting an optimal energy and materials mix in steel production.

Gaps in understanding the pathways of industrial decarbonisation

Despite substantial progress in the past, there are several research gaps in the literature on decarbonisation. First, there are insufficient studies on firms' behaviour on technological development and emissions control in the steel industry. Existing studies on emissions reduction in China's steel industry focus on carbon emissions quantification, emissions trading policies and on global comparisons in carbon emissions intensity for crude steel production. Technological options are commonly explored, as is emissions performance CCS progress, and low-carbon projects. But there has been no systematic investigation of how and why firms choose to adopt certain technologies and not others.

Second, existing studies of emissions in China's steel industry have neither considered the dual carbon strategy nor the changes of firms' behaviour in response to such strategy. More detailed studies on the temporal and spatial characteristics of pollutants emitted by China's steel industry often focus on non-carbon emissions. Recent projections of emissions, including a study on decarbonisation pathways, focus only on aggregate levels. Studies on emissions reduction potential in the steel industry, such as An et al., have not considered the implementation of the dual-carbon strategy.[6] Policy impact assessments have been conducted on standards, but not on the emissions-reduction or energy-saving policies that have been increasingly prevalent in the past decade.

Third, decarbonisation pathways for the Chinese steel industry towards zero carbon emissions will significantly impact on commodity markets and the Australian economy, but these matters have not yet been examined. China accounted for more than 80 per cent of Australia's iron ore exports and a quarter of its coking coal exports in 2020. A better understanding of the Chinese steel industry's decarbonisation pathways is necessary for Australia to understand the challenges and opportunities presented by future global commodity markets over the coming decades.

While Australia and China are closely linked in the steel industry, studies on the relationship are mostly outdated and mainly focus on supply chains. There are only a few studies on this relationship from an emissions perspective and quantitative assessments are notably lacking. For example, Zhang et al. (2021) examined the impact of a shift from iron ore export to direct reduced iron (DRI) exports from Australia to steel production in countries such as China, South Korea, and Japan, but did not consider other impacts beyond emissions reduction.[7]

Fourth, there is little understanding of the potential impact of digital technologies (industrial internet, artificial intelligence, big data and so on) on steel industry transformation along the pathways to decarbonisation. In 2017, Shougang Group announced the adoption of artificial intelligence on quality testing. Testing for 10,000 steel pictures revealed an accuracy rate of 99.98 per cent. Meanwhile, Baogang Group successfully lowered the tapping temperature by an average of 10 degrees through AI production control and saved 7 billion yuan on energy costs. Although there are successful cases of digital technology's application in steelmaking, there is limited cost–benefit analysis or assessment of long-term influences in the Chinese context.

Finally, with most countries having now set targets for reducing carbon emissions and reaching carbon neutrality, a significant question remains: to achieve agreed goals on limiting damage from climate change, will all countries have to reach carbon neutrality a decade and maybe fifteen years earlier than the current targets (2040 or 2035 for developed countries; 2050 or 2045 for China and Russia; 2060 instead of 2070 for India and other developing countries)? There is a big gap in our understanding of the programs for decarbonisation to see if the goals set by many countries, including China, can be achieved earlier than scheduled, as it has an important implication for achieving the global target of limiting the temperature rises for the twenty-first century and beyond.

Opportunities and potential

Moving along the pathways of decarbonisation represents the beginning of the decline of the traditional ways of industrialisation in China, which in the past have powered its industrialisation with high investment, resource and energy intensities, and high pollution. It also represents an opportunity for China to undertake the historical transformation of its economy towards more sustainable development. China must accomplish the task in a much shorter period than its international peers. The combination of measures discussed in this chapter – including structural reform, energy transformation, new technology and new patterns of international trade and wider cooperation – offers hope.

China's commitment to peak carbon emissions by 2030 and achieving carbon neutrality by 2060 requires a dramatic change in output structure, energy mix, regional and industrial location, and production technologies. China's decarbonisation strategy will go hand in hand with efforts to transform a factor-driven growth model into an innovation-driven one. Together with developments in renewable energy, infrastructure and the digital economy, this can result in continued economic growth with greater environmental amenity.

For Australia, China's decarbonisation as well as easing demand for steel in the next two to three decades will greatly reduce demand for iron ore and coking coal. But this can be replaced by new areas of cooperation that are larger and even more beneficial for both countries. The big energy and industrial transformation taking place in Australia and China suggests that there is a wide scope for both countries to cooperate in renewable areas, producing benefits for each other and for the global economy than those generated by earlier periods of enhanced economic cooperation between the two countries.

There is great potential for China's steel industry to be supplied processed materials embodying large amounts of renewable energy being produced in Australia. For producing iron, the natural path to decarbonisation is by replacing metallurgical coal by hydrogen made by splitting water with renewable electricity. Upgrading iron ore to semi-processed iron (or steel) exports is the key to that – with the iron content gradually falling as China uses more scrap and becomes less steel-intensive, but with the value of the trade being maintained or increased because of the higher value added. Of course, that would require a reset of the trade relationship, which is very much in the interests of both countries and of the world – including but not only what it does for the ease of decarbonisation in China.

Author's note

I appreciate the guidance and advice provided by Ross Garnaut in writing this chapter. I thank Rentao Rao for research assistance provided, and Yixiao Zhou for helpful comments and suggestions on an early draft. Any errors are mine.

7

LAND CARBON

Isabelle Grant

There is an old, proven, simple and very powerful tool to reduce global warming that sequesters carbon in vegetation as a negative emission technology (NET). This process, with a successful track record extending over several billion years, is photosynthesis. Photosynthesis converts solar energy and carbon dioxide from the atmosphere into carbon compounds in plants beneath and above the ground. After the plant's death, the carbon can be taken up by microbes and transferred to the soil. This technology will help Australia and the rest of the world keep warming to 1.5°C.

Total global carbon in soils is around 2500 GT,[1] compared to around 560 GT in living plants and animals and 760 GT in the atmosphere.[2] The high ratio of carbon in plants and soils relative to the atmosphere (9:2 in 2008) shows that a modest increase in landscape carbon could be associated with a large reduction in atmospheric carbon.

Approximately one-third of all global CO_2 emissions since 1850 have come from loss of carbon in terrestrial ecosystems; most of the remainder from burning fossil fuels.[3] Consequently, to limit global warming to 1.5°C, land carbon management and restoration is needed on a large scale to absorb historical emissions in the atmosphere and reduce emissions in the future. Achieving net-zero global emissions by 2050 will require substantial contributions of negative emissions from carbon in the landscape.[4] Australia can make a substantial contribution.

A range of expert assessments demonstrate potential opportunities for sequestering carbon in Australian soils and plants. A 2019 *Science* article showed that 205 GT of carbon could be sequestered by reforesting the world's woodlands and forests. Its authors calculated that Australia has the capacity for 58 million hectares of tree cover outside of major agricultural zones and areas dedicated to human development such as cities and towns. Australia was in the top five countries in the world in terms of potential to sequester carbon through reforestation. Australia contains 6.4 per cent of the 900 million hectares available for reforesting.[5] By comparison, the United States' contribution is three-quarters larger in absolute terms but one-tenth of Australia's per capita.

In *Superpower: Australia's Low-Carbon Opportunity* (2019), Ross Garnaut discussed the 2017 US Academy of Science study that demonstrated the global opportunity of a range of 'natural climate solutions'. It outlined 'twenty conservation, restoration, and improved land management actions that increase carbon storage and/or avoid greenhouse gas emissions across global forests, wetlands, grasslands, and agricultural lands'. This could sequester almost 24 GT of carbon per year, with Australia contributing 461 million tonnes.[6]

In fact, Australia's potential to sequester carbon and reduce emissions through natural climate solutions is even greater than this. Harnessing the large Australian landmass, naturally adapted plant species and new technologies for using bio-based materials in fuel, biochar and biochemicals, Australians could reduce global net emissions by almost 670 million tonnes of CO_2e every year for many decades. Of Australia's 770 million hectares, less than 10 per cent would be needed to make this change.

Annually, around 18 MT of CO_2 could come from planting mallee trees alongside highly modified crop and pastureland, 116 MT from regenerating woodlands and forests and 8.5 MT for every million hectares of trees planted in low rainfall zones across Australia. Of Australia's 770 million hectares of land, planting trees on 50 million hectares of low-rainfall areas could sequester 425 MT annually. As of 2021, 50 million tonnes of woody waste currently produced every year could reduce carbon dioxide emissions by about 38 MT a year if it were converted into sustainable aviation fuel. Planting of hedgerows in 1 per cent of all bare pastoral land would sequester 23 MT of CO_2e when used as a feedstock for biochar and bio-oil as a long-term carbon store and as a substitute for heating oil. Mixed species plantations could supply an extra 22.6 MT CO_2e as a biochar and bio-oil from 2 million hectares of agave and river red gum.

Forest and woodland regeneration potential

Carbon accumulates in plants through the growth of roots, stems, branches and trunks. It is deposited in the soil from the natural cycling of debris when a plant dies. There are many methods to sequester carbon in the growth of plants, including planting and nurturing trees. Natural regeneration is the cheapest in many circumstances. But it may not be the most cost-effective, nor will it necessarily lead to an optimal level of sequestration. Natural regeneration can be achieved by stopping grazing, agriculture and other forms of land use that suppress the growth of vegetation and by allowing the regrowth of native vegetation. Natural regeneration of parts of Australia to levels before European settlement would sequester carbon, and have many other benefits.

Older, larger trees contain much more carbon than smaller trees. They need to be prioritised during forest regrowth. When a forest grows, there are different stages of succession. Succession of temperate forests usually results in the dominance of one or two species and a reduction in the initial competition for space and light from other early-establishment species. Roxburgh et al. (2006) demonstrate the difference in carbon sequestration between managed regrowth of forests, compared to afforestation.[7] Unmanaged forest regrowth in historical plantation stands was achieving only 60 per cent of total potential carbon-carrying capacity. This reduced regrowth was mainly due to the absence of larger trees resulting from logging of the region. New studies since demonstrate the advantages of thinning forests to promote growth of older trees. Thinning is the process of cutting down denser stands of trees to make room for others. It gives the existing and new trees more space to grow. This suggests that there is greater carbon sequestration when trees grow at different rates and thinning increases structural diversity of stands.[8] As managed regrowth is a major component in the federal government's Emissions Reduction Fund (ERF), research into how we manage regeneration of native forests across Australia is important. The ERF provides carbon credits to projects that improve net carbon outcomes, such as regenerating forests. Our estimates of carbon sequestration and our potential sequestration from regenerating forests vary depending on how the regrowth process is managed.

The carbon sequestration potential of Australian reforestation has been measured in several studies. The US National Academy of Science estimated there to be 386 million tonnes of CO_2 per annum potential. Australian studies identify a higher current carbon stock than international generic estimates for corresponding climatic regions, or biomes. Mountain ash (*Eucalyptus regnans*) forests in Victoria hold the most carbon per hectare of any forests in the world.[9] The total carbon carrying capacity (a metric which represents a long-term carbon outlook) of southeast Australian forests shows that, at maturity, restoration of forests in New South Wales, Victoria, Queensland and Tasmania has around 33 billion tonnes of CO_2e potential.[10] That number is greater than the total cumulative emissions from Australia in the last hundred years. This staggeringly large figure would take up to 100 years to achieve.

Looking at the historical distribution of trees in these forests is a good place to start in estimating Australia-wide annual carbon sequestration potential from reforestation. In 1990, John Carnahan produced a vegetation map of Australia before European settlement. This map used a range of historical records, including environmental surveys, maps and notes, and compared these with satellite imagery of Australia in 1990. This historical map (Figure 7.1) can be compared to that of current Australia for an understanding of the loss of woodland and forest cover in the last 240 years. In 1788, there was an estimated 349 million hectares of forest[11] – 160 million hectares more than current estimates of forest and woodlands. Deforested areas are now covered by cropland, urbanised areas and pastures.

Default biomass (organic matter) figures per biome published by the Intergovernmental Panel on Climate Change (IPCC) in 2006 allow us to estimate potential annual increases in carbon. Releasing land from pastoral use, productive forests, old mine sites and unused land can support new forest areas where there were once native forests. In 1750, eucalypt woodlands were the second-largest vegetation group in terms of land cover. Only 36 per cent of these woodlands remained in 2016. Figure 7.1 shows that non-mallee eucalypt woodlands occur mainly in the eastern coastal regions and the north of the country, and spread over temperate, subtropical and tropical climate zones. Common eucalypts of northern Australia include Darwin woollybutt (*E. miniata*) and Darwin stringybark (*E. tetrodonta*). River red gum (*E. camaldulensis*) is the most widely distributed eucalypt and is found in all Australian mainland states. The forests of southeastern Australia contain a wide range of dominant eucalypt species, including major commercial timber species such as mountain ash (*E. regnans*), messmate stringybark (*E. obliqua*), alpine ash (*E. delegatensis*) and spotted gum (*Corymbia maculata*).[12] In temperate regions, we would see 4.4 tonnes of biomass per hectare per year in growth of new forests. Across tropical regions,

Figure 7.1 Total area of Australia by land vegetation type, pre-1750.

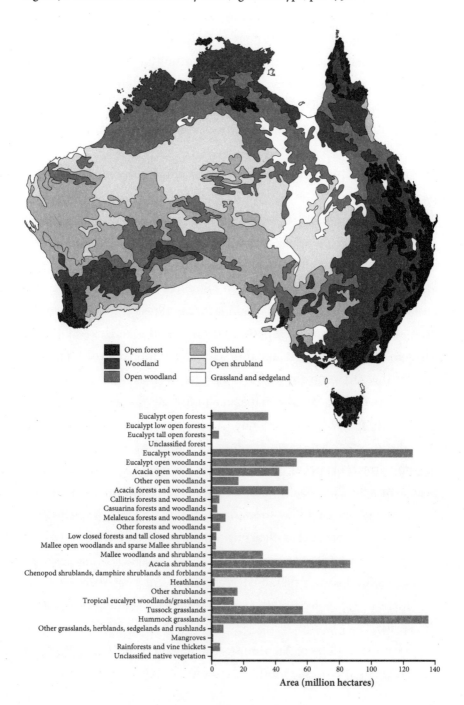

this would be 7 tonnes and in subtropical areas around 5 tonnes. Starting to restore 10 per cent of this woodland to its former distribution would provide 116 million tonnes of CO_2e sequestration a year.[13] As previously mentioned, sequestration opportunities per hectare are generally larger in the temperate forests of Australia than in similar climatic and soil conditions elsewhere and collection of site-specific data would provide us with a better estimate of these potential annual yields.[14]

Carbon opportunities for annual crops and highly modified pastures

This next section suggests how Australia could greatly expand native reforestation, with plantings outside the dark grey zones. Figure 7.2 shows Australia's current land uses, indicating where action could be taken to complement or transform areas with more tree cover. Planting trees on pasture and as plantations in marginal areas allows the areas in lighter grey and white in the middle of the map also be utilised. This could occur through planting native adapted species and xerophytic (dry-adapted) species that thrive in low and variable rainfall constraints, therefore reducing the need for water management in water-stressed regions.

It is not possible to regenerate all historical above-ground vegetation, due to urbanisation and land used for food production. Cessation of commercial harvesting of native forests has been an important means of reducing total national emissions since 2008. However, native woodland clearing rates in many states are outcompeting rates of restoration and there is concern that clearing is being underreported in federal accounts.[15] Around 54 per cent of Australia's landmass is managed for the agricultural and cattle industry, and in recent years there has been considerable focus on bringing land carbon management practices to this sector.[16]

Sixty-six million of Australia's 769 million hectares are used for annual crops and highly modified pastures (see Figure 7.2). Of these,

Figure 7.2 Vegetation in Australia by land use type, 2016.

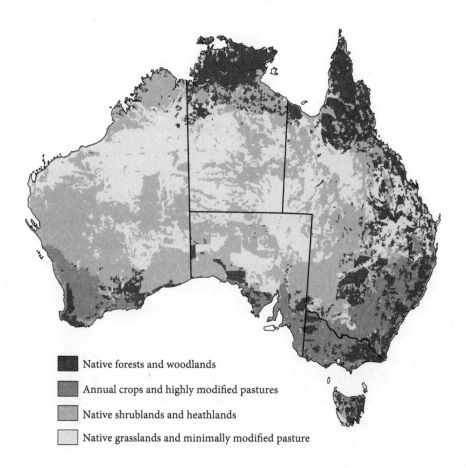

Native forests and woodlands

Annual crops and highly modified pastures

Native shrublands and heathlands

Native grasslands and minimally modified pasture

around 30 per cent are in mallee country,[17] characterised by a dry, flat and sandy soil and Mediterranean climate.[18] In Western Australia, this soil was converted from forest cover between the 1950s and 1970s, which has resulted in substantial salination of soil and subsequent agricultural losses. Mallees are the native species in the old forest eco-systems. Between 44 and 56 per cent of this old woodland still exists.[19]

Across Australia, if 10 per cent of the 20 million hectares of annual crops and highly modified pastures became mallee cover, this would total 18 million tonnes of CO_2 sequestration annually.[20]

A low-carbon economy in central west Queensland

To provide insight into the landscape carbon potential of Australia, we present a model for a low-carbon economy in the central west of Queensland. This area has unique flora, fauna and climate, but productivity is limited – as in many other areas – by water availability and rainfall variability. Water here is scarce and precious. Average rainfall is not low by Australian rural standards – between 350 mm and 500 mm per annum – but rainfall variability and recurring drought are big problems, leading to unreliability of vegetation growth constraining economic opportunity. This model demonstrates the growth of biomass from xerophytes (arid-adapted plants), and from the use of native species that are naturally adapted to the changing weather and fires of central Australia. Plants are elaborately adapted to their variable climate, and these adaptations can be undervalued in regions where classic forms of farming and forestry prefer fast-growing and high-yielding species.

Larger trees generally indicate a larger carbon sequestration potential, as size (biomass) is directly linked to quantity of carbon. However, native Australian acacia are naturally better at sequestering carbon irrespective of size, due to their water-use efficiency (WUE) and nitrogen-fixing ability. They can store a unit of carbon with less water and no nitrogenous fertiliser. This adaptation gives acacia a natural advantage in areas of low soil nitrogen and low water availability, as in marginal farming regions of Australia. The ratio of carbon to nitrogen in acacia is lower than that for other species, at around 12:1 or even 15:1, vs 30:1 in other tree species. In recent papers and a 2016 study, Mark Adams of Swinburne University and other researchers discuss the advantages of Australia's acacias for storing carbon in variable rainfall zones.[21] Reforestation opportunities in the Brigalow Belt and the mulga forests of Queensland and New South Wales would contribute greatly to Australian carbon sequestration.

Acacia can be grown in plantation-type systems with higher and faster biomass yields resulting from rotations. They can also be grown in intercropping, mixed species rows and silvopastoral systems, with value as shelter belts for animals. Rotations can sustain value. Establishment of saplings may require initial irrigation and some protection from feral dogs and kangaroos.

Another option is to use coppiced species. Coppicing retains the roots and base of the tree when the trunk is cut. The tree sprouts multiple stems for higher biomass yields. The mallee in the wheatbelts of Western Australia across southern Australia is highly suitable. Species endemic to northern and inland Australia such as *Atalaya hemiglauca* (commonly known as whitewood) and various acacia that are present across Australia, for example *A. victoriae, A. stenophylla* and *A. salicina,* can also be coppiced.[22] Coppicing was developed over 6500 years ago, in the Neolithic era in England, for firewood production. Its practice lasted until the 1800s, after which large-scale commercial operations took over. However, it has become more popular in recent years, supporting short rotation coppice (SRC) bioenergy systems in North America and Europe, using a range of species such as poplar (*Populus spp.*) and willow (*Salix spp.*).

While providing income in the form of carbon credits, carbon farming can also improve the productivity of crops and grazing. Agroforestry is the umbrella term for a range of tree-agricultural systems, whereby trees are placed on land to provide shelter belts for cattle, additional income from fruits or nuts, or to improve soil structure and nutrient cycling from leaf litter when deciduous trees drop their leaves in winter. One of the most well-known ancient examples is goat and cattle herding in the olive groves of the Mediterranean, a system called silvopasture.

The work of Dr Daniel Mendham and others at the CSIRO has contributed to a large base of literature demonstrating the benefits of

Figure 7.3. A silvopastoral system in Australia.

Source: Sellar Farmhouse Creamery.

trees in agricultural systems.[23] One paper calculated returns using a simple cost–benefit model from rows of trees on windy strips of land parcels used for grazing in Tasmania.[24] This model shows that rows of *Pinus radiata* plantings have a positive net benefit on the grazing of sheep, as the shelter decreases sheep mortality and the drying of pastures. Case studies using the Imagine bioeconomic model in Tasmania reveal reduced wind speeds and evaporation. These impacts have economic benefits. These case studies, seen all over the world, demonstrate that carbon is part of a bigger story reconnecting farmers with natural systems.

Agave is adapted to the extreme heat and long periods without rain of arid areas of central and western Mexico. There is extensive land available for plantings of *Agave tequilana*, as it grows naturally in

Figure 7.4 Eucalyptus and pasture – an example of agroforestry.

Source: cristhian9876 / iStock.

temperatures above 38°C. Desert areas and colder regions which reach temperatures below zero at night are not suitable for agave. Very large areas of Australia that currently have no agricultural and low pastoral value would be suitable for agave. These include many areas in range-lands and central Australia, such as large areas of Queensland, the middle of the Northern Territory and large areas of Western Australia. One million hectares of agave could lead to sequestration of about 4.6 million tonnes per annum of CO_2 in biochar, as well as the production of 6.6 million tonnes of bio-oil for energy or as an industrial input.

An article published in *Science* in 2019 – 'Forest Restoration: Overlooked constraints' shows that large areas of arid Australia face issues of 'salinity, sodicity, hard-pans and moisture limitations'.[25] It is important to invest in research on plant breeding and genetics to prepare for the increasing impact of climate change on these bioregions. Selection of species for their drought resilience, the sourcing of non-native drought-tolerant species and understanding the biochemical mechanisms that underpin these characteristics will help Australia sequester carbon in semi-arid and arid zones.

Also noteworthy is that climate change has already begun to impact Australia since European settlement in 1780 and reforestation opportunities in 2022 are not the same as in 1780. Degradation of the soil through intensive farming, thus reducing its fertility and ability to host forested areas, has no doubt reduced the potential reforestable area.[26]

Limitations of the large-scale plantation figures presented in this report such as variations across environmental gradients, lack of spatial in-situ tree data and climate change impacts should be recognised. Due to the marginality of the climate in large areas of Australia, the risk of prolonged drought and wildfires make the permanence of large scale plantations in these regions unpredictable. Plantations for marginal land have been investigated in the CSIRO report *Opportunities for Carbon Forestry in Australia: Economic assessment and constraints to implementation*[27] and in the 2008 RIRDC commissioned report *Regional Opportunities for Agroforestry Systems in Australia*.[28] The CSIRO provides estimates of total potential carbon sequestration for land that has been previously cleared, which overlaps with areas suitable for low-intensity plantations in Queensland. The RIRDC report describes plant species distribution across Australia's five main geoclimatic zones, including Northern savanna and semi-arid (>275 mm rainfall) and Southern (275–550 mm). No large-scale figures have been provided for plantings across these distribution maps. Opportunities for plantations in marginal areas are under-represented in both reports because of a lack of available data for plant growth in these regions. More research into possible large-scale vegetation growth models across marginal Australia is needed to realise the potential carbon sequestration here.

Biochar as a negative emissions technology (NET)

Production of a high-carbon composite called biochar is a NET that complements growth of plants for carbon capture. Trees can be harvested for pyrolysis once the plant has surpassed its maximum growth rate and can be replaced by new trees. Pyrolysis is the thermochemical decomposition of organic matter in an inert environment – that is, in the absence of oxygen. It produces two products: biochar – which is high carbon – and a syngas, or bio-oil.

Use of biochar dates back at least to the black soils, or *Terra preta*, of the Amazon basin. These soils were cultivated to compensate for low-fertility soils by the indigenous people of the Amazon basin, mainly in Brazil, but also in Peru, Ecuador, French Guinea and Bolivia. There were similar practices in places in Africa, between 450 BCE and 950 CE. The productivity in these areas is significantly greater than that of the surrounding lower-value soil.[29] *Terra preta* earth was created through incidental or purposeful low-temperature pyrolysis in nomadic and agrarian settlements. The origin of *Terra preta* is debated but is thought to have come from slash-and-char agricultural practices leading to high-carbon anthrosol soils. There is also archaeological evidence that ancient cooking practices had some contribution to *Terra preta*. The addition of charcoal to soils thousands of years ago has remained as *Terra preta* today and even increased carbon in soils since then.

The biochemical similarities in *Terra preta* and biochar, and their interactions with soil microbes and beneficial exchange with soil nutrients and minerals, present an interesting insight into the potential for biochar as a soil amendment and carbon store. Black soils today contain up to 15 per cent soil organic carbon content.[30] For comparison, dryland agricultural soils in Western Australia hold between 0.7 and 4 per cent soil organic carbon (SOC) in soils. In Australia, higher carbon is usually found in organic soils or soils in intensive dairy farming.[31]

Substantial scientific work and data on biochar corroborates this important long-term empirical data.[32] The IPCC's 2018 land sector mitigation report recommends biochar as a NET that will be important to limiting global warming to 1.5°C.

Research over the past decade has discovered that biochar fed to ruminant animals in feedlots can reduce methane emissions while increasing animal productivity by improving digestion of plant materials. Emissions reductions of 10 to 20 per cent have been reported.[33] More research is required to understand the possible methane reduction of ruminant animals in extensive grazing conditions. It also reduces methane in rice paddy fields when applied directly to soils.[34]

Biochar can have multiple roles in Australian agricultural and pastoral systems. Fed as a food supplement to ruminant animals (cattle and sheep), it reduces methane emissions and increases animal productivity. Returned to the ground as manure, it improves soil fertility and moisture retention. With certification that it has been returned to the soil, it sequesters the carbon reliably and promotes the growth of soil biota that can increase the carbon stock over time.

The 2018 IPCC report also reviews the literature on soil carbon sequestration techniques and their costs. The report looks at the net costs and benefits from animal productivity, soil fertility and the sale of co-products. The cost of soil carbon sequestration ranges from –US$45 per tonne of CO_2e to US$100 per tonne of CO_2e. Negative figures indicate benefits exceeding costs through co-benefits from trees and regenerating soils. Net biochar costs range from US$30 per tonne of CO_2e to US$120 per tonne of CO_2e. The cost of biochar compares favourably with the European carbon price, €87.50 (US$97).[35] The IPCC notes that the need for negative emissions technologies will increase as the carbon price rises over time along the path to zero net emissions. Alternative negative emission technologies are very expensive: advanced weathering (up to US$3640 per tonne of CO_2e) and ocean alkalinisation (US\$14 to >US\$500 per tonne of CO_2e).[36] It is economically desirable for the role of lower-cost solutions available now, such as using biochar as a carbon store, planting trees on farms and reforestation of native forests, wetlands and mangroves, to expand (avoiding the need for higher-cost and higher-risk alternatives).

The second product from pyrolysis is a syngas or bio-oil which can be burned for energy, upgraded as a road fuel or higher value bioproduct or input into zero-emissions manufactured goods to replace fossil carbon and hydrocarbon. Bioproducts from bio-oil include platform chemicals, bioplastics and resins.

Industrial uses of bio-oil and biomass

Bio-oil from pyrolysis can be upgraded to biofuels for transport, used as heat and electricity generation as an oil, and transformed into a range of bioproducts or platform chemicals. Biomass is extremely flexible to transformation into new products. The new bioeconomy will provide renewable substitutes for fossil fuels used as sources of carbon and hydrocarbon in manufacturing. The European Union predicts that this change could save 2.5 billion tonnes of CO_2e for the European community by 2030.[37] The EU biobased and industrial biotechnology industry was expected to reach \$1 trillion in 2022.[38]

Until now, research and industry for bio-oil transformation has been centred in Asia, Europe and North America. Nearly all European countries have published bioeconomy strategies. In 2019, 0.4 per cent of global bioplastic production was in Australia.[39] Australia's rich endowment of land compared with population and local demand provides us with an opportunity to play a much larger role.

Pyrolysis oil has been used for centuries as methanol, acetic acid and turpentine. Over 300 compounds are found in the oil, contributing to a range of industrial uses.[40] These compounds can be isolated and extracted or the pure bio-oil can be transformed through chemical reactions with other compounds. Aside from its uses for energy and heat, bio-oil can be used in making de-icers; slow-release fertilisers; pyrolytic lignin as resins and adhesives; levoglucosenone for pharmaceuticals; food flavouring; and biodegradable polymers and surfactants.

In the United States, the greatest consumption of petroleum and gas is in gasoline, distillate fuel oil and then hydrocarbon gas liquids (HGLs). HGLs contain ethane, propane and refinery olefins, all of which contribute to plastic, solvents, paint and synthetic rubber production. Bio-oil has been used in making three of these groups of products. Other forms of biomass have flexibility to produce these too, particularly in the biodegradable form. Figure 7.5 shows the wide range of bio-based plastic (biodegradable and degradable) production capacity in 2019.

Bio-oil could replace a range of petroleum products and fuels. An Australian spin-off of the chemical engineering department at Curtin University, Renergi Pty Ltd, is building its first commercial-scale pyrolysis project at Collie in Western Australia, to convert municipal waste and waste biomass into char and bio-oil. Using such technology, 1 million hectares of marginal rangelands plantations in northern Australia could produce 17 million tonnes of bio-oil if transformed from river red gum, or about 13.3 million tonnes if transformed from agave. Two million hectares of mixed agave and river red gum plantations across marginal Australian land could supply over 30 million tonnes of bio-oil over five and ten years, respectively. These 2 million hectares would supply 10.6 per cent of combined Japanese and South Korean oil demand.[41] Once established, annual rotations would save around 11 million tonnes of CO_2e from substitution of fossil-fuel petroleum consumption. The same biomass would sequester 7.6 million tonnes of CO_2e from the rotation byproduct of this bio-oil production in pyrolysis.

Biofuel can complement electrification and more efficient aeronautical design in decarbonisation of air travel. Short trips are more likely to be battery- or hydrogen-powered given the lower weight of energy storage

Figure 7.5 Industries' capacities of bio-based materials in 2019.

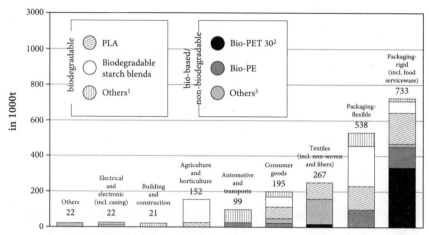

¹ Contains regenerated cellulose and biodegradable cellulose ester
² Bio-based content amounts to 30%
³ Contains durable starch blends, Bio-PC, Bio-TPE, Bio-PUR (except thermosets), Bio-PA, PTT

Source: Institute of Bioplastics and Biocomposites, Biopolymers, facts and statistics 2020, Hochschuhe Hannover, Germany, Edition 7, 2020.

required. Longer flights are constrained more by weight restrictions. Biofuel has lower-energy density than petroleum fuels but is likely to play a major role in the move to zero emissions in long-distance civil aviation.

Sugarcane residues on the Queensland coast, forestry industry residues in Victoria, South Australia and New South Wales and the agriculture and wine industry waste in Western Australia and South Australia could provide a large quantity of biomass for biofuel production. Some of the agricultural waste that comes from industry across Australia is burnt, unnecessarily returning carbon to the atmosphere. The 2021 Australian Biomass for Bioenergy Assessment identified over 50 million tonnes of plant waste annually.

Biofuel conversion would be centred around hubs and nodes linked to biomass sources by efficient transport systems. Through rural and provincial Australia, this would support new employment in urban centres and higher income from land use.

With the right logistical and technological planning, bioenergy hubs could produce the equivalent of around 13.5 million tonnes of sustainable jet fuel a year from waste residues alone. This is about 6 per cent of global demand.[42] Per litre of jet fuel, 3.3 kilograms of CO_2e[43] are released during combustion. Replacing jet fuel from oil with renewable biofuels produced from waste in Australia could remove around 38 million tonnes of CO_2e a year[44] – equivalent to almost 0.1 per cent of global and 6 per cent of Australian emissions.[45]

There are a wide range of potential sources of biomass for conversion into valuable products in Australia. Pastoral hedgerows of 0.1 per cent of all rangelands country in Australia would provide around 27.5 million tonnes of CO_2e per year left as environmental plantings and sequestering carbon over time. Conversion of hedgerows into biochar and bio-oil production through pyrolysis would provide about 23 millions tonnes of annual CO_2e savings under a ten-year rotation system with replanting after harvest of river red gums (*E. camalulensis*).

Soil carbon

Soils are becoming a major focus of land carbon sequestration policies in Australia and globally. The accumulation of soil organic carbon (SOC) is affected by a range of biotic and abiotic factors. The living (biotic) component consists of fungi, bacteria and invertebrates that change the structure and availability of nutrients in the soil. They break down plant and animal matter, and make carbon, nitrogen, phosphorus and many other minerals available to the plants and animals that live there. The abiotic (non-living) component is the minerals and nutrients, rock and soil types that make up the physical matter. The carbon:nitrogen (C:N) ratio has a strong influence on availability of SOC.

More vegetation provides greater availability of carbon through nutrient and carbon cycling. This is true for a great range of soils in Australia. Climatic and environmental changes can affect this process.

Drought has taught us that additional and more consistent rain is the main factor that will increase soil carbon through native vegetation cover. Planting of legumes is another. Legumes are nitrogen-fixing plants, due to the nodules on their roots that host certain types of bacteria. These bacteria can convert nitrogen in the atmosphere to nitrates, which can be taken up by the plant. Nitrates are a form of nitrogen plants use to build proteins to grow and acquire more biomass. These are essential for increasing the quantity and rate of carbon uptake in the plant, with the ideal C:N ratio for microbial activity around 24:1. When dead matter decomposition is reduced, less carbon is available in the soil. This reduces the mass of plants and invertebrates in the soil, causing a negative effect on soil fertility.

Legumes are one of the Conservation Agriculture pillars, which also include no-till and rotation grazing or farming. Conservation Agriculture is a global movement to improve farm sustainability and recognise the limits to traditional farming, particularly across semi-arid zones. Australia has been pioneering one pillar of Conservation Agriculture for longer than most, which is no- or reduced-till. Since the 1970s it has been recognised that the soils of the semi-arid areas of Australia suffered greater than others under tillage. The movement of the soils reduced their structuring, released carbon through aeration and undermined erosion-mitigation of the soils. Tillage allows precious topsoil to be blown away, leaving lower-value soil below, without the high-nutrient humus that characterises good soil health.

Blue carbon

Through tackling biodiversity, we can reduce unnecessary release of carbon and sequester carbon in plants and animals. It has been calculated that returning whale populations to healthy, sustainable levels would sequester as much as 2 billion mature trees.[46] This is because of their impact on the livelihood of phytoplankton. These phytoplankton absorb carbon via photosynthesis in huge quantities: 40 per cent of all CO_2 produced, at 37 billion tonnes of CO_2e, which is four times the size of the Amazon rainforest. The whales act as a 'pump' for key minerals, particularly in warmer parts of the ocean, that are required for phytoplankton growth. Through migrations, whales deposit nitrogen, iron and phosphorus in the epipelagic layer of the ocean, which hosts great quantities of phytoplankton.

Australia is just recognising the gains in carbon it can obtain from changes to the way we manage coastal and freshwater ecosystems. Blue carbon projects can absorb up to four times more carbon per hectare than rainforests and continue to sequester carbon longer than trees on land.[47] Australian carbon credits for blue carbon have only just been made available.

Total blue carbon in Australia is estimated to sequester 20 million tonnes of carbon per year.[48] Australia's tidal marshes, which cover 1.4 million hectares of land, store 212 million tonnes of carbon at an annual rate of 0.75 million tonnes per year.[49] 133 million tonnes of carbon are stored in macroalgae across Australia, but rising sea temperatures put this at risk, causing ocean acidification, macroalgal disease and changes in marine communities that feed on macroalgae.[50] Protection and repopulation of these essential blue carbon stores could unlock huge value across Australia.

Biodiversity concerns

To understand humankind's impact on the planet, the planet's resilience and tipping points that threaten Earth's integrity, the Stockholm Resilience Centre research the limits of nine planetary boundaries. 'Rockstrom's planetary boundaries' tell us we are exceeding our planet's ability to manage with this loss of biodiversity.[51] As supporting whale populations in the wild can increase carbon, scientists are now looking into reintroduction or 'trophic re-wilding' of herbivores and predators that also have large impacts on the quantity of carbon in the landscape. *Superpower* noted that efficient harvesting and processing of kangaroos could increase value and reduce carbon emissions from meat production. Rewilding initiatives such as these would include a concerted effort from many players, such as government, conservation groups and landholders, to implement at scale.

Other opportunities to increase carbon across the landscape involve increasing biodiversity. Managing weeds can increase the diversity of native plant species and even return native invertebrates and birds to the region through habitat creation. These invertebrates can be beneficial to re-establishing biological control of pests on farms. Across vineyards, cotton, brassica, peanut and cereal crops proximity to native vegetation has demonstrated high abundance of crop pest predators and natural predation of pests.[52] The native vegetation typically carries a low abundance of the pests themselves, therefore providing a low-risk, low-cost alternative to pesticides, including for species such as the green ant (*Oecophylla smaragdina*), which is a natural pest for over fifty insect species.[53] This wide-ranging biological pesticide provides as strong protection against pests as some chemical pesticides. Native predation of pests from vegetation demonstrates one of many ecosystem services that plants provide alongside carbon.

The Australian carbon price

Global awareness of nature-based solutions for climate change has led to substantial investment in restoring tropical rainforests, restoration of high-carbon peat- and grasslands and large-scale carbon plantations across the world. Markets for carbon in Europe, parts of North America and in voluntary global markets provide incentives for communities to plant trees. As of April 2022, forty-six national and thirty-six subnational jurisdictions have established national carbon pricing mechanisms to ensure they keep within the nationally determined contributions (NDCs) laid out in the Paris Agreement.[54] These contributions outline pathways to reducing emissions and detail extensive land carbon sequestration. With Article 6 of the Paris Agreement made operational at Glasgow in 2021, trade in credits will become more important in complying with targets. Australia has a limited market for carbon offsets, based on the Australian Energy Regulator's Emissions Reduction Fund and safeguard mechanism, and is in the early stages of linking this into national and global carbon trading systems.

Since *Superpower: Australia's Low-carbon Opportunity* was published in 2019, the price of carbon in the limited Australian market has increased considerably. From a stable price around $16 over the previous twelve months, it increased to $54.50 in November 2021. The carbon price currently sits around $35. The rise was partly driven by global and national pressures in the lead-up to the Glasgow conference and following it. Most importantly, voluntary acceptance of emissions reduction goals by major companies led to increased demand. Federal support for new carbon sequestration methodologies such as lower-cost soil carbon measurement and carbon capture and storage was announced in 2021, as well as a record project registration per quarter with the Climate Solutions Fund, which administers carbon credit projects in Australia.[55] These increases in supply of credits were overwhelmed by increasing demand. Demand increased in February 2021 in response to safeguard mechanism breaches: major emitters including Alcoa and Anglo American had to buy more credits.[56] The price continued to increase until an extraordinary disruption of market arrangements by the federal government in March 2022 (Figure 7.6). There was a sudden and unexplained release of contractual requirements to deliver Australian carbon credit units (ACCUs) to the Commonwealth government within the Emissions Reduction Fund. This reduced the price of carbon and calls the future reliability of Commonwealth-level Australian carbon pricing into question. This followed the Abbott government's

Figure 7.6 Spot price (AUD) of ACCUs from late 2021 to mid-2022.

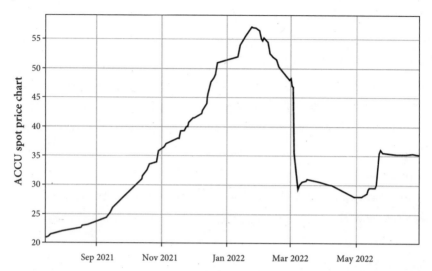

reduction in the Renewable Energy Target in 2015 and the abolition of national carbon trading in June 2014. Together, these three changes in established arrangements constitute the most damaging manifestation of sovereign risk in Australia in living memory.

Problems in the Australian market

1. Integrity

The integrity of ACCUs became a subject of scrutiny in Australia in early 2022 – particularly credits administered for regeneration projects that appear to have undersupplied carbon against CSF predictions. ANU law professor Andrew Macintosh and colleagues highlighted the need for more comprehensive measurement of real carbon sequestration.[57] The Australian carbon accounting system has largely received domestic support so far, but the Australian government must respond positively to the Macintosh critique if local and international support is to be retained. The highest number of ACCUs registered (over 50 per cent of all contracts so far) is in the human-induced regeneration (HIR) method. This is one of the two methods that have been called into question by Professor Macintosh.

2. Lost opportunity

There is an absence of close correlation between carbon sequestration outcomes and credits allowed under narrow and sometimes arbitrarily defined methodologies. Single methodologies are only allowed per property, meaning that gains in carbon through other methods cannot be awarded to the landholder. Development of an integrated technique is underway under the ERF and will aim to solve some problems of lost opportunity.

3. Cost

High costs of measurement, particularly in soil carbon measurement, curtails the potentially large contribution that landscape carbon in Australia can make to containment of increases in atmospheric concentrations of carbon.

It is necessary to move away from partial approaches to measurement of changes in carbon stocks, to more comprehensive accounting for augmentation and depletion. This requires more cost-effective, reliable approaches to measuring carbon in soils and plants.

Measurement of land carbon: Remote sensing

The global community has come together to expand the world's ability to observe changes in carbon from space using satellites, lidar (light detection and ranging) and radar. Australia has excellent capacity to do this for its own accounting system.

Three international satellite missions are combining their efforts over the next couple of years. The NASA GEDI mission uses NASA lidar and ICESat (ice, cloud and elevation satellites) to assess land emissions through 3D imaging. It was the first of the satellite missions to launch, in 2018. The second mission is the NASA–ISRO (NISAR) Indian radar satellite mission launching in 2023. The third is the European Space Agency Biomass radar mission, which also launches in 2023. These missions, with various benefits for calculating high-resolution changes to vegetation from satellites, will work together to improve the world's understanding of carbon in the landscape.

The Terrestrial Ecosystem Research Network (TERN), Australia's land ecosystem observatory, has been monitoring the changes in soil carbon emissions from sites across Australia. It uses an atmospheric measurement technology called Eddy Co-variance Flux Towers, which measure CO_2 and water vapour changes in the atmosphere to predict the release and uptake of CO_2 as it happens. These towers provide a great baseline of information to understand regular changes in CO_2 fluxes for bioregions in Australia. The Queensland University of Technology is compiling data on soil carbon changes across several sites in Queensland that have new management practices to assess changes to soil carbon. TERN is working to improve the remote sensing of soil carbon from satellites and drones. When measurement of these data from satellites becomes sufficiently reliable, it will complete the cycle for remote sensing of carbon in plants.

Rainbow carbon accounting: An introduction

These technologies have the potential to form the basis of a new accounting system for landscape carbon. In *Superpower* and *Reset*, this was called comprehensive carbon accounting. The work of Indigenous leader Suzanne Thompson of the Iningai people in central west Queensland has led us to prefer the term rainbow carbon accounting (RCA), which covers the whole range of notional colours of sequestration. RCA would record all changes to carbon stock on national land and coastal accounts. It would be a comprehensive system.

What does rainbow carbon accounting mean for the Iningai people in Queensland?

[The land is] formally known as Gracevale Station and now known locally as "Turraburra" after the traditional clan that fell to the push inland by colonial occupiers. Turraburra has become a showcase for traditional land management practices and a demonstration model for Rainbow Carbon values and cultural enterprise.

What YACHATDAC (Yambangku Aboriginal Cultural Heritage and Tourism Development Aboriginal Corporation) is defining as rainbow carbon has evolved from the application of indigenous land management practices to the QLD Land Restoration Project. A snapshot of Indigenous land management practices and comprehensive nature-based solutions for carbon sequestration, abatement and regenerative environmental actions include the following:

- Revegetation of endemic species: planting and harvesting of traditional bush foods and medicines
- Drought proofing
- Wildfire protection
- Revitalisation of previously existing canopy vegetation (HIR method)
- Traditional fire stick mosaic and cool burns (soil sequestration)
- Rotational livestock grazing and reduction of herd sizes
- Restoration of natural springs and wetlands
- Revitalisation of waterholes and waterways
- Land erosion management
- Ceremonial and wild harvesting practices
- Maintenance and preservation of cultural landscape and cultural heritage assets.

YACHATDAC's comprehensive approach to managing climate change with rainbow carbon and nature-based environmental solutions is an opportunity for the national and global communities to encourage and participate with Indigenous relations who are implementing traditional land management principles.

Source: S. Thompson, Email to Isabelle Grant, 25 April 2022.

The list of practices at Turraburra (outlined in the box above) highlights the range of colours in the RCA spectrum. For example, land erosion management represents brown carbon; traditional fire stick mosaic and cool burns represents red carbon; and so on. By seeing the full spectrum of 'colour' in the opportunity to revegetate the landscape, we can move away from rigid methods to sequester carbon in the landscape and focus on measurement, however it is sequestered.

A new accounting system

In monitoring and verifying carbon projects, the increase in carbon must be:

- additional
- unique and traceable
- permanent
- real and measurable
- independently verifiable.

These requirements are well known. Some major problems in practice are:

- permanence
- additionality
- double-counting
- carbon leakage.

RCA would make the carbon dimension of the outcomes from improved management of the landscape operational for purposes of carbon markets. It would also avoid increasing problems with the limited methodologies of the ERF. It could be accompanied by explicit recognition of values beyond carbon in biodiversity, conservation and Indigenous heritage, which could attract separately accounted credits.

Existing methods for monitoring changes in carbon in the landscape in national accounts would be bolstered and eventually replaced by a technology that can detect real-time changes at the landholder level. Participation by landholders would be voluntary. Participants would be comprehensively rewarded for increases and penalised for reductions in carbon stocks. Initial participation on a pilot basis would provide information to assess systems. Initial participants should cover a range of bioregions and climate environments and a suite of remote sensing technologies.

The formation of such a system would require accelerated development of measurement technologies. Development of remote sensing for measuring carbon in soils features in the previous Australian government's Technology Roadmap. This needs to be extended to measurement of carbon stocks above the ground. Several Australian universities and the CSIRO are in the final stages of testing and commercialising technologies for large-scale remote sensing of carbon. Lidar technology has become increasingly sophisticated in recent years and long-term work at CSIRO and the University of Melbourne has seen models for measuring carbon in above-ground vegetation developed. Lidar detects distances from the lidar scanner with a laser using light reflection. By obtaining many distance points this builds a visual 3D picture, defining the size and shape of the object or landscape. Size and shape are then correlated to carbon sequestered.

The cost of measurement is a deduction against the returns from carbon farming. Rewards for carbon should be readily available at reasonable cost wherever activities are genuinely contributing to an increase in carbon sequestration.

RCA pays participants for all increases in carbon, no matter the method. Landholders hold themselves accountable to managing carbon as they are penalised for reductions and rewarded for accretion. Because most of the carbon is measured remotely from satellite or drones, the cost of participation is lower than under the old methodologies.

In carbon credits 'additionality' is a difficult thing to prove. It is a requirement that is there to prevent credits being given to projects that were already going to happen. For example, planned industry changes, such as the development of organic farming, longer plantation rotations and technical improvements that reduce emissions and cut operational costs. Additionality requires that offsets are rewarded for a project only when that decision would not have occurred without the carbon offsetting mechanism or policy. Additionality is difficult to prove when we are attempting to assess the motivations or actions of a landholder.

The most important new rules for RCA will be around baselines, averaging and financing. Baselines for carbon projects in plantation forestry are currently assessed by the common industry practices of the time. There are legal, financial and technical assumptions that form baseline scenarios. For improved plantation projects, the following are examples of those considered for baselines under the Californian cap-and-trade system: 'forest growth and mortality, forest practice rules, best management practices, deed restrictions and other legally binding documents, timber harvest plans, logging costs, generated timber revenue, inaccessible areas'.[58] Monitoring carbon from satellite and remote sensing removes the emphasis on legal and financial constraints. Ongoing, constant measurement of carbon in the atmosphere, vegetation and soil will inform baseline-related rules and guide through difficult questions on the additionality of carbon.

Baselines are corruptible if they aren't carefully selected to be universal and administered in a timely manner. Baselines should be taken on the year of RCA announcement. This will reduce the risk of artificial degradation of carbon to earn credits from regrowth. The participant is required to confirm they have not artificially degraded land to achieve carbon credits, and this will be monitored regularly and can be looked at retrospectively.

The current carbon sequestration monitoring, reporting and validation (MRV) involves forecasting models for vegetation growth and ground truthing. These models are built from extensive data on growth of a huge range of native species and plantation growth models. The accuracy of these models is assessed periodically and updated with new data and research. This model is very well resourced and has proven to be successful at predicting carbon sequestration for particular methodologies. The benefit of methodologies for land carbon has guided landholders on what to plant, where to plant and the costs and effectiveness of different management practices. The existing methodologies will still work under an RCA operation, but RCA opens the possibility for other practices to demonstrate genuine increase carbon stocks.

Insurance will address concerns for landholders at high risk from accidental losses in carbon for example through environmental degradation from rising temperatures and extreme weather events, particularly bushfires. Because this system will reward landholders for carbon sequestered and also penalise them for losses, the landholder is liable for changes. It is in their best interests to protect and promote activities that sequester carbon; this will inevitably foster a positive carbon outcome for participants. However, in cases of loss of carbon for any reason – including landclearing and mismanagement of vegetation regrowth, for example, as well as misfortune – payments will be required back to the governing body. There will be important contractual agreements that tie participants to these losses, such as land title as security.

Above all, this new system will support new carbon farming projects by monitoring and sensing, instead of forecasting and modelling.

Models will help understand expected changes to inform decision-making until we have real results to prove these changes. Landholders will be kept accountable for changes and will prioritise land management that provides positive carbon sequestering outcomes.

8

THE RENAISSANCE OF RURAL AND REGIONAL AUSTRALIA

Susannah Powell

Rural and regional Australia stand to benefit the most from Australia's low-carbon opportunity, through new jobs and direct investment. The opportunities for competitive production of zero-emissions industrial products are disproportionately located in country Australia. The renewable energy, mineral and biomass resources for manufacturing zero-carbon metals, other processed critical minerals, processed biomass as inputs into zero-emissions manufactured goods and zero-emissions chemical manufactures such as fertiliser are located away from major cities.

Seizing these opportunities will lead to the renaissance of rural and regional Australia and to an expansion of employment, incomes and population far greater than in the days when Australia rode on the sheep's back.

Opportunities in traditional coal areas
The rural areas with advantages for the earliest and lowest-cost start

include the coal regions, where electricity transmission infrastructure was established to take power from generators based in or near coal-fields to the cities and other locations. Old transmission and industrial infrastructure can be given new purpose: bringing renewable energy from good locations to new industrial centres near the locations of the old coal-based generators. These include the Latrobe Valley in Victoria (and the transmission linkage across Melbourne and Geelong to the industrial city of Portland in the southwest); the Hunter Valley and Illawarra in New South Wales; the Upper Spencer Gulf in South Australia; central Queensland, extending from the coastal industrial city of Gladstone west through the coalfields of the Bowen Basin to the rich underdeveloped coal resources of the Galilee Basin, where there were once high hopes, now disappointed, of a cornucopia of jobs from coalmining and power generation; and Collie in Western Australia.

In central Queensland, for example, the proximity of world-class solar and good wind resources, mountain slopes and depleted mine pits for storing energy, reliable access to water and land for biomass and transmission networks originally developed for coal is unusual in Australia and the world. High rainfall and large flows in the rivers provide water for growing biomass and industry, with opportunities to use in new ways the water flows and infrastructure once dedicated to coal and gas mining.

Similar advantages, but without the established transmission built for coal generators, are present in north Queensland. The north has the additional advantage of proximity to exceptionally rich resources of critical minerals that are increasing in value with movement to zero emissions. The ports and industrial cities on the Queensland coast from Gladstone in the south through Rockhampton and Mackay to Towns-ville provide centres of commercial and industrial strength without close parallels in Australia. The flat country west of the Great Dividing Range between the latitudes of Townsville and Gladstone, outside the cyclonic influences that make solar plants risky, represents the richest solar region connected by high-voltage transmission to Australia's National Energy Market. Wind resources are of high quality through the Great Dividing Range, especially on the western side.

Breaking the transmission constraint in Queensland

The combinations of exceptionally high-quality renewable energy and energy storage and biomass resources of these regions are concentrated within a rhombus-shaped region stretching from Townsville to Gladstone along the coast, and inland about 600 km to Barcaldine in the southwest and Hughenden in the northwest. In a submission to the Queensland government, the World Wildlife Fund described this as a 'Rhombus of Reliability' for zero-emissions development. The Queensland government's Bradfield Regional Assessment and Development Panel report identified the Rhombus of Reliability as the location of globally important zero-emissions economic activity, supporting large-scale expansion of employment and population.

Minerals processing

Townsville, connected by established transport infrastructure to rich mineral resources, has the opportunity to build processing for local minerals for which demand is growing rapidly to meet the requirements of the world's energy and industrial adjustment to zero emissions. These local minerals include copper, zinc, manganese and aluminium – as well as what the Quad heads of government meeting in September 2021 called critical minerals – commodities required for new energy generation and storage, including silicon, cobalt, magnesium, nickel, lithium, vanadium, graphite, titanium and the rare earths.

Green hydrogen and manufacturing industry

Future supply chains in the net-zero world will rely on green hydrogen produced using renewable energy in many industrial processes. With hydrogen necessary for production of ammonia for fertilisers and explosives, production of ammonia derivatives such as urea naturally gravitates towards low-cost renewable energy centres in Australia. Here the biggest opportunity is use of renewable hydrogen for converting iron ore into iron metal. The largest opportunities will emerge around the massive ore deposits in the Pilbara region of Western Australia. But the new iron-processing industry will do well first in smaller ore deposits close to established infrastructure and industrial skills in many decentralised locations.

Biomass for industry

In addition to abundant solar and wind resources, Australia's high per-capita water and land availability provide natural advantage in the production of biomass for industry and also for sequestration of carbon. Away from the coast, where rainfall is lighter and more variable, vast areas can generate substantial volumes of biomass or carbon credits through use of the plant species adapted to high temperatures and variable rainfall. Biomass is a source of zero-emissions industrial inputs and energy, whether used directly or through the use of bio-oil or biogas from pyrolysis. Biomass or bio-oil from its conversion through pyrolysis can be used as an industrial input replacing coal, oil or gas. Converted into liquid fuels, it can have especially high value in long-distance civil aviation, for which low ratios of energy to mass make batteries and hydrogen fuels cells unsuitable. Biomass can provide an essential input for production of zero-emissions nitrogenous fertilisers. The use of carbon dioxide derived from combustion of bioenergy with oxygen waste from electrolysis allows for zero-emissions conversion of ammonia into urea.

Biosequestration in living plants and in soils from biomass growth or productive plantations, combined with other methods of drawing carbon into the land (the application of biochar to soils, regenerative land management and reforestation), also present a significant opportunity across large areas in regional Queensland. While the Australian compliance market has been artificially repressed by unstable and poorly thought-out Commonwealth government policies since 2013, international arbitrage and expectations are providing upward price pressure, and carbon drawdown will continue to increase in value. This opportunity is explored in Chapter 7 in more detail.

Renewable energy precincts: A model for regions in transition

The Barcaldine Renewable Energy Zone (the BREZ), located in the centre of the Galilee Basin coal and gas region, takes advantage of Queensland's world-class renewable energy resources and potential for biomass production. The BREZ is designed to pioneer sophisticated zero-emissions industry manufacturing, including several different pyrolysis technologies for converting biomass and wastes into valuable char, oils and gases; the use of zero-emissions hydrogen for ammonia and urea; the processing of minerals; and intensive horticulture. The activities outlined in this chapter which are currently being explored for implementation would generate local investment (in 2021 dollars) of $1.8 billion over ten years, and up to $3.9 billion over twenty years. They would secure around 500 new permanent jobs across ten new businesses and 1000 construction jobs.

The BREZ model can be applied to many other locations in rural and regional Australia. Barcaldine was chosen as the first site to focus on due to the world-class renewable energy opportunity, the established supporting infrastructure and the engaged community. Once demonstrated, this model can be replicated with variations for local circumstances across regional Queensland to meet the growing global demand for zero-carbon minerals processing and agricultural commodities, utilising low-cost renewable energy and hydrogen.

Barcaldine energy resources

The renewable energy resources in the Barcaldine region are abundant. The solar resource is the best of all locations connected to the national electricity grid in eastern Australia and is outstanding by international standards. The wind resource is good.

With an existing 132-kilovolt transmission line connecting to the NEM via Clermont, Barcaldine can use the grid to balance the intermittency of local renewable energy resources. The scale of the balancing is limited by the capacity of the transmission line, but is sufficient to support industry large enough to transform the economic fortunes of the local community. Barcaldine is far enough inland to enjoy high insolation and to avoid the impact of tropical cyclones on renewable energy infrastructure. Further west, the solar insolation rises slightly but lacks grid connection. There are excellent wind sites immediately east of Barcaldine, along the west-facing slopes of the Great Dividing Range.

The combination of the extremely high-quality solar resource with good-quality wind resource facilitates consistent supply, as illustrated

Figure 8.1 Spring average daily energy production.

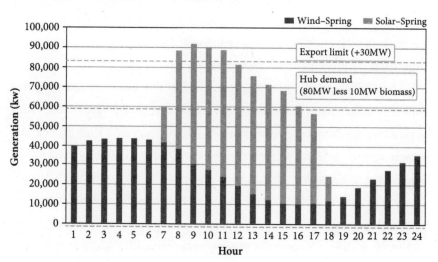

in Figure 8.1. The profile of the wind resource lends itself favourably to complementing solar generation. For the BREZ, it means a significant proportion of energy demand can be supplied by blending local wind and solar supply.

Part of the biomass feedstock resource will initially be supplied by noxious prickly acacia, which has degraded 22 million hectares of pastoral land across northern and central Queensland. Using prickly acacia will contribute to its eradication. As prickly acacia supply is exhausted, feedstocks will be provided from productive plantations as outlined in Chapter 7.

Using the same pyrolytic conversion technology, bio-oil can also be supplied by municipal waste from shires further east in central Queensland, including the Central Highlands around Emerald, Isaac with its headquarters in Moranbah, and the larger coastal cities of Mackay and Rockhampton.

Barcaldine water and mineral resources

Barcaldine has a long-term average annual precipitation of 491 mm.[1] This is not low by the standards of much of Australia but is highly variable. The standard climate models suggest that global warming will increase average rainfall but make it more variable in inland areas near and north of the Tropic of Capricorn.

Located along the northeastern part of the Great Artesian Basin (GAB) on the Alice River, the BREZ will draw water sustainably from a number of aquifers within the GAB. Within established regulatory arrangements administered by the Queensland Department of Regional Development, Manufacturing and Water, the BREZ will purchase water rights from private and public entities. The BREZ managers are exploring with private landowners and regulatory officials the capping of free-flowing bores on private land. Part of the old water flows is being made available for alternative uses. Capping free-flowing bores saves significant quantities of water that are currently being lost from the GAB to seepage and evaporation. Working within the GAB regulatory structures will ensure that demand for water in the BREZ avoids negative impact on the Great Artesian Basin – and probably has positive effects. Water will be made available at reasonable overall costs.

Barcaldine is located within the Eromanga Basin, on local sedimentary bedrock from the Mesozoic Era. To the northwest, the world-renowned Proterozoic-aged Mount Isa mineral province has immense prospectivity for many critical minerals, with large undeveloped resources of copper and cobalt. Within the Eromanga Basin, vanadium deposits within the organic-rich shale and limestones of the Toolebuc formation around Julia Creek are being developed by Vecco Group to source ore for its proposed processing plant to be housed at the BREZ. The Toolebuc formation outcrops along the eastern margin of the Eromanga Basin all the way south from Julia Creek to Barcaldine, providing potential for local sources of ore.

Market opportunities

The BREZ model, with its low-cost resource advantages, has significant market opportunities in energy-intensive sectors where synergies can be realised in green chemical supply chains. The co-location of production facilities enable byproducts and wastes of one process to be provided as low-cost inputs for complementary industrial processes. Exploiting such synergies is at the heart of the BREZ. Examples include the use of waste oxygen from hydrogen electrolysis as an input for oxy-firing of bio-oil in a steam turbine to produce zero-emissions thermal power and carbon dioxide for conversion of ammonia into urea. Waste heat energy, carbon dioxide, pure carbon and biochar from pyrolysis units will be inputs for intensive horticulture to maintain glasshouse climates and nutrients for plant growth. Biomass waste from the production of charcoal and native essences and flavours will be a pyrolysis feedstock. Processing of metals will use globally competitive zero-carbon energy. For example, at the BREZ site the electrolyser plant that turns vanadium into a base product for flow batteries will use energy as the primary input cost for zero-emissions manufacturing.

Vanadium batteries have a longer despatch life than lithium-ion batteries, and while less responsive are suited to industrial loads and will play an important part in global low-emissions energy systems.

The market opportunities for the production of green hydrogen, ammonia and urea are expanding and the BREZ access to resources makes it a good location for production at globally competitive prices. Urea is one of the top three globally traded chemicals. It is an important fertiliser in Australia, which imports almost 92 per cent of around 2 million tonnes per annum domestic demand. Many Australian crop farmers rely on urea as their primary fertiliser source. It is also used as a stockfeed supplement (including in central Queensland), and for other industrial applications such as the diesel fluid additive AdBlue. Conventional urea production uses fossil gas or coal as a feedstock and is highly emissions-intensive. China dominates global coal-urea production, primarily for domestic consumption, but also as a swing exporter. Russia and other Black Sea states and the Middle East dominate global exports. All introduce elements of supply chain insecurity into Australian agricultural and pastoral activities. With much of the international trade sourced from Middle East fossil gas producers, the cost of conventional, high-emissions urea has been driven to unprecedented heights by the various pressures lifting global gas and coal prices. The urea market is characterised by high volatility driven by seasonal weather patterns and global shipping demand.

Over the five years to 2020, international prices were in the range of $340 to $640 (in today's dollars) per tonne. Longer-term average prices in ports of origin fluctuated around $400 per tonne, with occasional higher spikes. Shipping to Queensland typically added $50 to the landed cost (in 2022, it is much higher). Transport to inland agricultural and pastoral regions often added a similar layer of costs.

Prices in Australia have risen sharply over the past year, reflecting high gas prices, steep rises in freight costs and reluctance to invest globally in new capacity in highly emissions-intensive activities. China's reduction of exports of this energy-intensive product in implementation of commitment to domestic decarbonisation has also contributed to the price rise, which has since been exacerbated by disruption of sales from Black Sea ports resulting from the Russian invasion of Ukraine. Some farmers in the central west and southwest of Queensland, including the major demand centres in the Darling Downs, have reported difficulties in securing supplies over the past two years. Prices more than doubled in that time. In 2022, urea prices have exceeded $1400 per tonne.[2]

The urea bill of one grazier in the Barcaldine region grew from $1 million to $3 million as a result of the price increases in 2021–22. For small to medium-sized food producers and large broadacre farms, the cost plus uncertainty about supply inhibits food production.

The impact has been felt beyond Queensland. Tasmanian-based former McCain grower representative and seed potato grower Beau Gooch says 'the fertiliser costs have put a double-figure increase into the cost of growing spuds'.[3]

While price varies and current price extremes are temporary, global constraints on expansion of emissions-intensive production from fossil-fuel (coal and gas) sources are expected to place continuing upward pressure on international prices and to generate ongoing concerns about supply-chain reliability. As China attempts to reduce gas and coal use in domestic manufacturing to meet climate change objectives, large urea exports are unlikely to return.

Secure access to urea at a stable price is important to sustain the prosperity of Australia's agricultural production. Secure urea supply is also crucial for domestic freight (trucking).

The zero-carbon urea plant located at the BREZ would be the first of its kind in Australia – and the world. Early studies show that it can produce green urea for $600 at point of production. At this price the Australian-made urea can be expected to be competitive with imports on average for the foreseeable future, with import prices rising as decarbonisation imperatives affect world costs and prices. Zero-emissions urea would strengthen the access of Australian farm and station products in international markets – at first with the developed countries leading the global decarbonisation effort and seeking zero-emissions supply chains for food and industrial inputs, and then more generally. Local production would clearly increase domestic supply security. In addition to an inherent freight cost advantage, domestic supply delinked from gas input prices and the vagaries of international shipping would reduce the exposure of Australian farmers to highly volatile commodity price fluctuations and security concerns. It is an economically rational aspiration for Australia to shift from being a large net importer of urea to a net exporter over the next one or two decades. This would see the basis of pricing to domestic farmers shifting from import parity (prices in overseas places of production – mostly the Middle East and Black Sea – plus freight) to export parity (international prices less freight). This is potentially a downward swing of $100 per tonne to Australian farmers at current international freight rates, and more in inland areas away from the import nodes.

Within the greater Barcaldine region, urea has a strong market – it has a high value in the cattle feed market, where it is used in high-protein molasses-rich supplements. This is a natural first market. A second local market is the Central Highlands nitrogenous fertiliser market.

Conventional urea production from coal and gas, as at the Incitec Pivot plant in Brisbane and in all the international suppliers to Australia, is one of the most emissions-intensive industrial processes. This exposes agriculture users to significant carbon risk. This risk is likely to mature into penalties in Europe from 2026, and in other developed countries soon after.

Most emissions from conventional urea production arise in the production of hydrogen from steam reformation of fossil hydrocarbons. These emissions are removed by making hydrogen from electrolysis using renewable energy. There are also emissions from energy used in the production of ammonia from hydrogen and urea from ammonia, unless renewable electricity is used. Carbon dioxide is used to convert ammonia (made from hydrogen) into urea. This absorbs carbon dioxide. If the carbon dioxide has a renewable source, the whole production process has substantial negative emissions. The BREZ would host the world's first zero-emissions commercial production of ammonia, and also the first conversion of ammonia to urea using biogenic carbon dioxide (the pure stream of carbon dioxide from combustion of bio-oil in a steam turbine for power generation using oxygen waste from the production of hydrogen through electrolysis). It would turn one of the most carbon-intensive elements of the agricultural supply chain into a source of negative emissions. This would provide competitive advantage for Australian farmers in high-value Australian and international markets. It would provide Australian farms with protection against future restrictions on emissions-intensive imports. It would also be a positive differentiator and a source of a premium for low-carbon produce.

Products with certified low- or zero-emissions supply chains would have an advantage in market access. Beyond carbon penalties implemented by our trading partners or Australia's national policy, demand for low-carbon commodities is being driven by multinationals such as Microsoft, Kellogg's, Procter & Gamble and McDonald's, which are implementing net zero targets throughout their supply chains. Due to shareholder and consumer pressure, corporate targets are more aggressive than government targets, with net zero by 2030 a common goal.

Urea is a compelling example of a market opportunity for sophisticated manufacturing in rural and provincial Australia. The principles of competitive advantage due to zero-carbon energy resources, supply chain optimisation from co-location of complementary industrial processes (pyrolytic conversion of waste into valuable inputs such as waste heat, pure carbon, bio-oil and char) apply across the products manufactured at the BREZ.

Building industrial precincts

If the rich renewable energy resources of rural and regional Australia are to be used to support new economic activity and employment, it is necessary to develop services in a coordinated way. Integrated precincts, such as the BREZ, are more important in new and isolated industrial regions than in large, established cities. In an established industrial city, a wide range of services and industrial inputs are available through competitive market processes, by incremental expansion of established systems. In new industrial regions, mechanisms must be found to coordinate many decisions on investment and provision of services through the early stages of development.

Development in rural Queensland faces a number of cost disadvantages that are only likely to be overcome by using the immense potential for low-cost electricity and biomass. Unlocking this potential requires the development of decentralised industrial precincts and the use of rural Queensland's immense land and water resources to produce low-cost electricity.

Building decentralised precincts of diverse industrial activity, with access to globally competitive power and good infrastructure services, can overcome the disadvantages of small scale and isolation from major centres of industrial activity by facilitating access to globally competitive power and good infrastructure services.

State governments can make an important contribution to the

establishment of new industrial precincts in rural and regional Australia by supporting initial funding of local studies on a small number of Sustainable Development Precincts, based on access to established transport, electricity transmission and other infrastructure, location in relation to biomass, renewable energy and minerals resources, other aspects of economic opportunity, and local interest. Barcaldine is the first of several potentially important developments in rural Queensland. Longreach, Hughenden, Emerald and several possible sites in the Darling Downs and Moranbah in the Bowen Basin are promising candidates for evaluation as possible rural industrial precincts.

The development of renewable energy infrastructure and regulation is critical to rural Queensland realising this economic opportunity. Barcaldine has been included as one of the Queensland Renewable Energy Zones. This may facilitate timely and cost-effective connection of the BREZ and the wind and solar farms to the Ergon network centre and the high-voltage transmission line joining Barcaldine to the main grid near Clermont.

A second and larger step would include a major expansion of the transmission line from Barcaldine to the coast, as part of building the Rhombus of Reliability in central and north Queensland. It is important that the Queensland government policy agencies and government energy business enterprises examine the role that substantial upgrading of the transmission connection from Barcaldine to the main grid could have in underwriting industrial growth along the route and especially at Gladstone. This should be done within a larger evaluation of upgrading of high-voltage transmission in the rhombus-shaped area joining Gladstone, Barcaldine, Hughenden and Townsville. Gladstone and Townsville are joined by substantial network connections linking these towns to Rockhampton and Mackay along the coast, and the towns through the coalfields of the Bowen Basin and eastern Galilee Basin. There are current plans for strengthening transmission between

Hughenden and Townsville as part of the Copper String project connecting Townsville to the mining region centred on Mount Isa in northwest Queensland. The connection of Hughenden would complete the Rhombus of Reliability, underpinning zero-emissions industrial development in Gladstone, Rockhampton, Mackay, Townsville, Hughenden, Barcaldine and the large area of central and north Queensland within its boundaries.

While there is immense potential for renewable energy generation at low cost at various points within the Rhombus of Reliability, and for using the local landscape for hydroelectric storage and hydroelectricity generation to firm wind and solar power, the current transmission hosting capacity severely limits its use. This is already a critical issue for the six identified Renewable Energy Zones in north and central Queensland, where the existing transmission can support less than 6 per cent of the potential generation.

AEMO's least-cost development pathway suggests 2 GW of additional utility stage energy storage will be required in Queensland in the next decade (and more than 4.5 GW out to 2040). This makes no allowance for utilising Queensland's large opportunities to supply zero-emissions processed minerals and other products, including hydrogen derivatives, to world markets. Currently only 350 MW is committed, including the 250 MW Kidston Pumped Storage Hydro Project. Several other storage proposals are at preliminary stages of development, including the 1000 MW Borumba Dam Pumped Hydro project. Arbitrage opportunities in the Queensland electricity market indicate storage is already economically viable, and its further development will both support renewable energy developments and ensure Queensland Electricity prices remain competitive. Electrolysers to produce zero-emissions hydrogen contribute positively to power security and reliability with large-scale use of renewable energy for industry if regulatory arrangements are designed to recognise their value.

Figure 8.2 The Rhombus of Reliability.

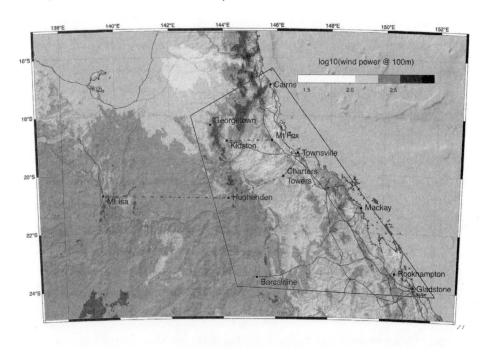

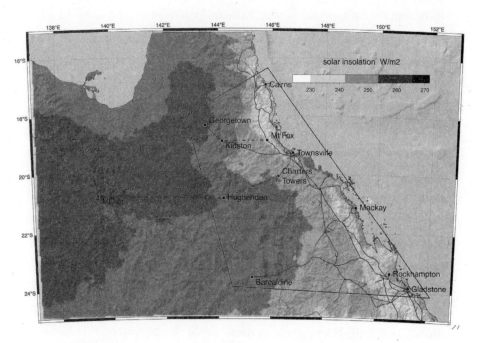

Resolving the transmission and storage constraints and associated energy cost disadvantage will enable the precinct model to deliver on the key priorities of the state government: to 'power regional development' and 'grow our regions [by] helping Queensland's regions grow by attracting people, talent and investment, and driving sustainable economic prosperity'.[4] It also will contribute to the Queensland government's ambitions: 'to become a renewable energy superpower',[5] develop Renewable Energy Zones and reach its targets of 50 per cent renewable energy by 2030 and zero-net carbon emissions by 2050.

The BREZ opportunity

After years of stagnation in much of rural and regional Australia, reform and infrastructure to support the BREZ model (in combination with the natural resource base) would create new jobs, and generate economic and population growth. It would provide a foundation for rural Queensland to be a global leader in the production of high-value agricultural commodities and in zero-carbon minerals processing. The Barcaldine model could be replicated across rural Queensland and in other parts of the country.

Barcaldine councillor Rob Chandler, who was mayor from 2004 to 2020, recalls the incredible hype around the Galilee Basin coal prospects between 2008 and 2014. In the early 2010s, the Queensland government gave environmental approvals for six mines in the Galilee – including Carmichael. Anticipating a need for housing for mine workers, Barcaldine Regional Council developed forty residential blocks between 2009 and 2010 in an adjacent town. The first sold for an average of $111,000. A year later, ten sold at $80,000. The council still holds twenty residential blocks and would not be expected to make $10,000 if they were put on the market in mid-2022.

Sean Dillon, the current mayor of Barcaldine and owner of a cattle station north of Alpha that sits over one of the proposed mines, says that genuine hope within communities was generated by the promises of jobs and income. Communities have accepted that there is little sign of the coalmines delivering the prosperity promised with many projects shelved, or discontinued as lenders and investors have withdrawn from financing new thermal coal projects.

People in the regions are looking for economic diversification. With climate change and the Paris Agreement commitments taken into consideration, global capital is seeking projects that have low or zero carbon risk. The Barcaldine model is a sustainable pathway to long-term prosperity and economic diversification for regional areas, returning them to their former role as Australia's economic engine.

ENDNOTES

Abbreviations

ABS = Australian Bureau of Statistics

ACCC = Australian Competition & Consumer Commission

AEMC = Australian Energy Market Commission

AEMO = Australian Energy Market Operator

IEA = International Energy Agency

IRENA = International Renewable Energy Agency

RBA = Reserve Bank of Australia

1 *The Bridge to the Superpower*

1   Ross Garnaut, with David Llewellyn Smith, *The Great Crash of 2008*, MUP, Carlton, 2009.

2   Ross Garnaut, *Economic Ideas and Policy Outcomes: Applications to climate and energy*, FH Gruen lecture, 29 June 2022; forthcoming in *Asian-Pacific Economic Literature*.

3   Ross Garnaut, *Dog Days*, Black Inc., Melbourne, 2013, p. 245.

4   Ross Garnaut & Anthony Clunies Ross, *Taxation of Mineral Rents*, Clarendon Press, Oxford, 1983.

5   Ross Garnaut, *The Garnaut Climate Change Review: Final Report*, Cambridge University Press, Port Melbourne, 2008, Box 14.5, pp. 346–7.

6   Andrew Macintosh et al., 'Integrity and the human-induced regeneration method: The additionality problem explained, ANU, Canberra, July 2022; Andrew Macintosh & Donald Butler, Response to Emissions Reduction Assurance Committee, June 2022.

2 *The Diminishing Carbon Budget and Australia's Contributionto Limit Climate Change*

1   IRENA, 'Renewable Power Generation Costs in 2021', July 2022.

2   S. Chowdhury, U. Sumita, A. Islam & I. Bedja, 'Importance of policy for energy system transformation: Diffusion of PV technology in Japan and Germany', *Energy Policy* 68, 2014, pp. 285–93.

3 Friedlingstein et al., 'Global Carbon Budget 2021', accessed at https://essd.copernicus.org/articles/14/1917/2022/, 23 July 2022.

4 IEA, 'Global CO_2 emissions rebounded to their highest level in history in 2021', press release, 8 March 2022.

5 Liebreich Associates, 'How to save the planet: Be nice, retaliatory, forgiving & clear', 17 September 2012, accessed at www.liebreich.com/how-to-save-the-planet-be-nice-retaliatory-forgiving-clear/, 24 July 2022.

6 'India may miss 2022 solar target of 100 GW by 27%', *PV Magazine*, 13 April 2022.

7 Gt CO_2 eq are billion tonnes of CO_2 equivalence, using the global warming potential (GWP) with a 100-year time horizon, provided by the IPCC AR6 WG1 report, to express non-CO_2 greenhouse gasses in CO_2 equivalence terms.

8 UNFCCC Secretariat GST Synthesis report, March 2022, available at: https://unfccc.int/documents/461517.

9 See Figure 2b in Malte Meinshausen et al., 'Realization of Paris Agreement pledges may limit warming just below 2 °C', *Nature* 604, 2022, pp. 304–9.

10 The business-as-usual baseline used for those countries with 2030 NDC targets that include far higher than projected emissions is based on downscaled emissions from SSP5 (J. Gütschow, M.L. Jeffery, A. Günther & M. Meinshausen, 'Country-resolved combined emission and socio-economic pathways based on the Representative Concentration Pathway (RCP) and Shared Socio-Economic Pathway (SSP) scenarios', *Earth System Science Data* 13(3), pp. 1005–40.). This SSP5 reference scenario is considered a less and less likely outcome at the global level (Z. Hausfather & G.P. Peters, 'Emissions–the 'business as usual' story is misleading', *Nature*, 2020, pp. 618–20). However, we use the SSP5 reference scenario in this context because of the considerable uncertainty in downscaled scenarios and because carbon leakage could put upward pressure on emissions by moving emission-intensive activities to countries and sectors not covered by an NDC. This makes it appropriate to err on the side of a higher emissions reference scenario for this baseline.

11 Figure 2b in Meinshausen et al., 'Realization of Paris Agreement pledges'.

12 This is in line with the SSP2-4.5 scenario explored by the global climate science and modelling communities and the IPCC (AR6 WG1).

13 IEA, *Net Zero by 2050*, flagship report, May 2021, available at: www.iea.org/reports/net-zero-by-2050.

14 IEA, *World Energy Outlook 2021: Technical note on the emissions and temperature implications of COP26 pledges*, IEA, 2021, available at: https://iea.blob.core.windows.net/assets/aa17bd09-2ad0-4d0a-b5aa-ee418900c4af/Theimpactsofnewemissionspledgesonlongtermtemperatures.pdf.

15   IEA, *World Energy Outlook 2021: Technical note.*

16   Figure 6 in Meinshausen et al., 'Realization of Paris Agreement pledges'.

17   Long-lived timber products (for example, in buildings) might lock away carbon in harvested wood products for the lifetime of a building, but this is not typically the many hundreds of years required of a biospheric store.

18   Several of the long-term targets are expressed in terms of GHGs, so there is some uncertainty about the implied CO_2 emission levels. We estimate CO_2 emissions reductions under these long term targets by approximating shares of gases across similar multi-gas scenarios.

19   'Five effects of the COP26: The Paris Agreement becomes more effective', https://blog.oeko.de/fuenf-ergebnisse-der-cop26-das-pariser-abkommen-wird-wirkungsvoller/#english, accessed 24 July 2022.

20   For background, the United States put forward very similar methane-focused alliances at least twice in the previous two decades, the Global Methane Initiative in 2004 and other initiatives focused on short-lived climate polluters over the years. Under the Bush administration in 2004, the Global Methane Initiative was viewed by some as an excuse for doing very little on fossil CO_2 and under the Kyoto Protocol. Often, a strong focus on shortlived climate forcers is proposed to buy time for the magical technological solution to fossil CO_2 to emerge. Yet, it is CO_2 that accumulates in the atmosphere, so any delay on CO_2 will cost us for centuries. See 'Five effects of the COP26'.

21   European Commission, 'Launch by United States, the European Union, and partners of the Global Methane Pledge to keep 1.5C within reach', statement, Brussels, 2 November 2021.

22   Council of the European Union, 'Council agrees on the Carbon Border Adjustment Mechanism (CBAM)', press release, 15 March 2022.

23   H.D. Matthews & K. Caldeira, 'Stabilizing climate requires near-zero emissions', *Geophysical Research Letters* 35(4), 2018; S. Solomon et al., 'Irreversible climate change due to carbon dioxide emissions', *Proceedings of the National Academy of Sciences* 106(6), 2009, pp. 1704–9.

24   C.D. Jones et al., 'The Zero Emissions Commitment Model Intercomparison Project (ZECMIP) contribution to C4MIP: Quantifying committed climate changes following zero carbon emissions', *Geoscientific Model Development* 12, 2019, pp. 4375–85.

25   Table SPM.2 in IPCC WG1 AR6, Summary for Policy Makers, available at: www.ipcc.ch/report/ar6/wg1/.

26   The reason is that IPCC AR6 references the warming levels not against pre-industrial (~1750), but against 1850–1900, which is a difference in anthropogenic warming of ~0.1°C with wide uncertainty (–0.0°C to +0.2°C). This reference period was chosen for two main reasons: For one, we have

more certainty about global-mean temperatures since 1850 when instrumental records began. Another reason is that the computationally intensive climate models usually start to model the anthropogenic effect on the climate from 1850 onwards. Every 100 years modelling time is precious, and given that comparatively little happened before 1850, it is considered to be a less efficient use of computing time to start models from, say, 1750. That 'comparatively little' warming before 1850 might be zero, but the best estimate for the anthropogenically induced warming is actually 0.1°C.

27 The conclusion by IPCC AR6 WG1, published in August 2021, that we only have 500 $GtCO_2$ of cumulative carbon emissions remaining after 2020 is based on the assumption that extra non-CO_2 warming is relatively modest. The new IPCC WG3 report, released in April 2022, used the latest generation of multi-gas emission scenarios, and found that 500 $GtCO_2$ might be actually generous. It could be that other non-CO_2 emissions would bring additional warming, so that the budget should be 400 $GtCO_2$. Or alternatively stated, that a 500 $GtCO_2$ budget results in pathways that temporarily overshoot 1.5°C by up to +0.1°C. WG3 found that 1.5°C pathways with no or limited overshoot have on average 510 $GtCO_2$ cumulative emissions from 2020 onwards.

28 UNFCCC Secretariat GST Synthesis report, March 2022, available at: https://unfccc.int/documents/461517.

29 Adapted from Figure 5 in UNFCCC Secretariat GST Synthesis report.

30 In 2014, the New York Declaration on Forests pledged to halt deforestation by 2030, like the UN Forum on Forests in 2017. The Glasgow pledge similarly pledged to halt deforestation by 2030, see 'Global leaders pledge to end deforestation by 2030', *The New York Times*, 2 November 2021.

31 See, for example, https://www.climate-resource.com/tools/ndcs/countries/chn?version=.

32 See, for example, C. Trudinger & I. Enting, 'Comparison of formalisms for attributing responsibility for climate change: Non-linearities in the Brazilian Proposal approach', *Climatic Change* 68, 2005, pp. 67–99.

33 Decision 1/CP.21 on the Adoption of the Paris Agreement states in paragraph 52: 'Agrees that Article 8 of the Agreement does not involve or provide a basis for any liability or compensation', where Article 8 in the Paris Agreement is about loss and damage.

34 G.J.M. Phylipsen et al., 'A triptych sectoral approach to burden differentiation: GHG emissions in the European bubble', *Energy Policy* 26 (12), 1998, pp. 929–43.

35 See also www.paris-equity-check.org as an example of various mitigation approaches.

36 Bundesverfassungsgericht, 'Constitutional complaints against the Federal Climate Change Act partially successful', press release no. 31/2021, 29 April 2021.

37 The simplified illustrative calculation here is that a 1.5°C carbon budget of 500 Gt CO_2 from 2020 is assigned proportional to the 0.33 per cent population share to Australia, resulting in a CO_2-only budget of 1650 MT CO_2 from 2020 onwards. Australia's targets are not specified in CO_2-only terms, but applying the 26 per cent target below 2005 levels and a net-zero target by 2050 as if there were just CO_2, result in ~7050 MT CO_2 of emissions between 2020 and 2050. If the 2050 net-zero GHG target were implemented, actual CO_2 emissions would be lower, as net-negative emissions have to be implemented to offset remainder CH_4 and N_2O emissions from the agricultural sector. On the other hand, the 'plan' that was brought to Glasgow, did not actually envisage a net-zero GHG achievement by 2050. Hence, our illustrative comparison between 7050 MT CO_2 and the 1650 MT CO_2: that is, a factor of 4.3 might approximately paint a correct picture of the scale of the mismatch of Australia's target in terms of a 1.5°C ambition.

38 The modified contraction and convergence resulted in an equitable share of 0.97 per cent, whereas the factor that is merely a reflection of Australia's population share is 0.33 per cent. A factor of three difference.

39 Disclaimer: We contributed to the calculations of the Climate Targets panel, whose findings are available here: www.climatecollege.unimelb.edu.au/australias-paris-agreement-pathways.

40 Malte Meinshausen, Yann Rubiou du Pont & Anita Talberg, *Greenhouse Gas Emissions Budgets for Victoria*, briefing paper, Australian–German Climate & Energy College, University of Melbourne.

41 Ironically, the so-called Greenhouse Development Rights (GDR) or Climate Equity Reference approach, an approach favoured by some in the Climate Justice movement for its apparent inclusivity of a number of equity principles, yielded some of the highest emission allocations for Australia: 1.26 per cent. That is because it accepted uncritically as a starting point Australia's own weak 'business as usual' starting point. While the GDR or Climate Equity Reference approach played a large role in civil society discussions, it was never important in the negotiations.

42 This is done to enable a subsequent comparison of fair shares approaches for Australia that includes the Climate Change Authority (2014) estimate of 0.97 per cent based on modified contraction and conversion.

43 G. Grassi et al., 'Critical adjustment of land mitigation pathways for assessing countries' climate progress', *Nature Climate Change* 11, 2021, pp. 425–34.

3 *The Net-Zero Opportunity for Australian Minerals*

1 IEA, *The Role of Critical Minerals in Clean Energy Transitions*, Paris, 2021; Lukas Boer, Andrea Pescatori & Martin Stuermer, *Energy Transition Metals*, International Monetary Fund Working Paper, 12 October 2021.

2   IEA, *The Role of Critical Minerals in Clean Energy Transitions*.

3   'Global wind turbine fleet to consume over 5.5Mt of copper by 2028', Mining. com, 2 October 2019.

4   Guillaume Pitron, *The Rare Metals War*, Scribe, Melbourne, 2020, Kindle edition, p. 14.

5   Peter Farley, personal communication, June 2022.

6   The fifty mineral commodities proposed for inclusion on the 2021 list of critical minerals: aluminum, antimony, arsenic, barite, beryllium, bismuth, cerium, cesium, chromium, cobalt, dysprosium, erbium, europium, fluorspar, gadolinium, gallium, germanium, graphite, hafnium, holmium, indium, iridium, lanthanum, lithium, lutetium, magnesium, manganese, neodymium, nickel, niobium, palladium, platinum, praseodymium, rhodium, rubidium, ruthenium, samarium, scandium, tantalum, tellurium, terbium, thulium, tin, titanium, tungsten, vanadium, ytterbium, yttrium, zinc, and zirconium.

7   See: https://e360.yale.edu/features/china-wrestles-with-the-toxic-aftermath-of-rare-earth-mining.

8   Nicholas Niarchos, 'The dark side of Congo's cobalt rush', *The New Yorker*, 31 May 2021.

9   IEA, *The Role of Critical Minerals in Clean Energy Transitions*.

10 R.B. Gordon, M. Bertram & T.E. Graedel, 'Metal stocks and sustainability', *Proceedings of the National Academy of Sciences of the United States of America*, 103(5), pp. 1209–14.

11 Melissa Pistilli, 'Is peak copper coming?' *Copper Investing News*, 23 November 2021.

12 IEA, *The Role of Critical Minerals in Clean Energy Transitions*.

13 Boer, Pescatori & Stuermer, *Energy Transition Metals*.

14 Frik Els, 'Car makers will "need to become miners" – Benchmark' Mining. com, 1 April 2022.

15 Brian J. Skinner, 'Earth resources', *Proceedings of the National Academy of Sciences of the United States of America* 76(9), pp. 4212–17.

16 The 1980 wager between business Professor Julian Simon and biologist Paul Ehrlich. Ehrlich chose copper, chromium, nickel, tin and tungsten. The bet was formalised on 29 September 1980, with 29 September 1990 the payoff date. All five commodities that were bet on declined in price and Ehrlich thus lost the bet.

17 Julian Simon, *The Ultimate Resource II*, Princeton University Press, 1998.

18 RBA, 'Composition of the Australian economy: Snapshot', www.rba.gov.au/education/resources/snapshots/economy-composition-snapshot/, accessed 8 July 2022.

19   Geoscience Australia, 'Australia's identified mineral resources', www.ga.gov.au/scientific-topics/minerals/mineral-resources-and-advice/aimr, accessed 8 July 2022.

20   Nils Pratley, 'Dig it up and ship it out: How BHP keeps it simple', *The Sydney Morning Herald*, 16 January 2010.

21   ABS, *International Trade in Goods and Services, Australia*, Table 5, 5368.0.

22   Scope 3 iron ore processing in 2020 totals over 900 $MTCO_2$-e RioTinto 2020 (376.4 Mt). Rio Tinto, Scope 1, 2 and 3 emissions calculations methodology 2020, www.riotinto.com/-/media/Content/Documents/Invest/Reports/Climate-Change-reports/RT-climate-scope-123-report.pdf, accessed 8 July 2022; BHP (293 $MTCO_2$-e) BHP, 'Reducing our value chain emissions', www.bhp.com/sustainability/climate-change/reducing-our-value-chain-emissions, accessed 8 July 2022; Fortescue (246 $MTCO_2$-e) FY21 FMG Fortescue, 'Climate change and energy', www.fmgl.com.au/workingresponsibly/climate-change-and-energy, accessed 8 July 2022; Australian emissions (491 $MTCO_2$-e) 'Australia's greenhouse gas emissions: March 2021 quarterly update', Department of Industry, Science and Resources, 31 August 2021.

23   Geoscience Australia, 'Australia's identified mineral resources'.

24   Iluka Resources Limited, 'Eneabba rare earths refinery final investment decision', Australian Securities Exchange Notice, 3 April 2022.

25   Jarrod Lucas, 'WA minister gives green light to build $500m rare earths refinery at Kalgoorlie', ABC (online), 2 February 2022.

26   Paul Hunt, 'Building future Groote progress', *Australia's Mining Monthly* 27 May 2021.

27   Quentin Dehaine et al., 'Geometallurgy of cobalt ores: A review', *Minerals Engineering* 160, 2021, p. 106656.

28   Annastacia Palaszczuk & Scott Stewart, Joint statement, 'States first vanadium mine a new era for Queensland's resources sector', 15 September 2021.

4 *The Decarbonisation of Electricity*

1   Dick Warburton et al. (eds), *Renewable Energy Target Scheme: Report of the Expert Panel*, Commonwealth of Australia, Canberra, 15 August 2014.

2   Elouise Fowler, 'Rooftop solar growth expected to slow in 2022', *Australian Financial Review*, 23 February 2022.

3   AEMC, *Residential Electricity Price Trends Report 2021*, EPR0086, 2021.

4   ACCC, *Inquiry into the National Electricity Market – November 2021 Report*, Canberra, 2021.

5   Warburton et al., *Renewable Energy Target Scheme*.

6 AEMO, *Quarterly Energy Dynamics Q4 2021*, Melbourne.

7 Dylan McConnell & Mike Sandiford. *Winds of Change: An analysis of recent changes in the South Australian electricity market*, Melbourne Energy Institute, 2016.

8 ACCC, *Inquiry into the National Electricity Market*; AEMC, *Quarterly Energy Dynamics Q4 2021*.

9 Alinta Energy, 'Closure of Flinders Operations,' Alinta website, https://alintaenergy.com.au/about-us/power-generation/flinders-operations.

10 Engie., 'Hazelwood Power Station in Australia to close at the end of March 2017,' Engie website, 3 November 2016, www.engie.com/en/journalists/press-releases/hazelwood-power-station-australia, accessed 12 July 2022.

11 Kellie Lazzaro, 'Worksafe notices detail extent of repairs needed at Hazelwood' ABC News (online), 1 December 2016.

12 For more details, see Dylan McConnell & Mike Sandiford, 'Impacts of LNG export and market power on Australian electricity market dynamics, 2016–2019', Current Sustainable/Renewable Energy Reports, 7, 2020, pp. 176–85.

13 AGL, 'Liddell Innovation Project Frequently Asked Questions', AGL website, 2018, www.agl.com.au/-/media/aglmedia/documents/about-agl/sustainability/rehabilitation-and-transition/frequently-asked-questions.pdf, accessed 12 July 2022.

14 AEMO, 'NEM Electricity Statement of Opportunities 2021', AEMO website, 2021, https://aemo.com.au/en/energy-systems/electricity/national-electricity-market-nem/nem-forecasting-and-planning/forecasting-and-reliability/nem-electricity-statement-of-opportunities-esoo#:~:text=is%20available%20here-,2021%20Electricity%20Statement%20of%20Opportunities,and%20'green'%20hydrogen%20consumption, accessed 22 July 2022.

15 Origin Energy, 'Origin proposes to accelerate exit from coal-fired generation' Origin Energy website, 17 February 2022, www.originenergy.com.au/about/investors-media/origin-proposes-to-accelerate-exit-from-coal-fired-generation/, accessed 12 July 2022.

16 AEMO, *Quarterly Energy Dynamics Q4 2021*.

17 Clean Energy Council, *Clean Energy Australia Report 2022*, CEC, Melbourne, 2022.

18 Ross Garnaut, 'Catch the energy superpower tide to defeat recovery headwinds', *Australian Financial Review*, 12 December 2021.

19 The market operator is an independent non-for-profit corporation, charged with operating and planning the National Electricity Market.

20 This included publishing the National Transmission Network Development Plan (NTNDP) annually.

21 Alan Finkel et al., *Independent Review into the Future Security of the National Electricity Market: Final report*, 9 June 2017.

22 The ISP's purpose is 'to establish a whole-of-system plan for the efficient development of the power system that achieves power system needs for a planning horizon of at least 20 years for the long-term interests of the consumers of electricity.'

23 The inaugural plan was published in 2018, the second iteration in 2020 and the 2022 version is due to published mid-2022.

24 AEMO, *2022 Integrated System Plan: For the National Electricity Market*, Melbourne, June 2022.

25 For a detailed breakdown of the scenarios and input assumptions, see AEMO, *2020–21 Planning and Forecasting Consultation on Inputs, Assumptions and Scenarios*, 10 December 2021.

26 AEMO 2021c

27 Anthony Macdonald et al., 'Cannon-Brookes fronts $8b bid to buy and shut AGL coal', *Australian Financial Review*, 20 February 2022.

28 AEMO, *2022 Integrated System Plan*.

29 These actionable projects still need to be approved through the Regulatory Investment Test for Transmission, known as the RIT-T.

30 AEMO, *Addendum to the Draft 2022 ISP for the National Electricity Market*, Melbourne, 11 March 2022.

31 Peter Hannam, 'South-East Australia risks temporary gas shortages by 2023 winter, energy review warns', *Guardian Australia*, 29 March 2022.

32 Ross Garnaut, *Reset: Restoring Australia after the Pandemic Recession*, LTUP, Melbourne, 2021.

33 CEC, *Clean Energy Australia Report 2022*.

34 A recent ANU study suggests Australia could grow the electricity sector by a factor of twenty-seven: Paul J. Burke et al., 'Contributing to regional decarbonization: Australia's potential to supply zero-carbon commodities to the Asia-Pacific', *Energy* 248, 1 June 2022.

5 *Hydrogen*

1 Hydrogen use statistics are available in S. Griffiths et al. 'Industrial decarbon-ization via hydrogen: A critical and systematic review of developments, socio-technical systems and policy options', *Energy Research & Social Science*, 80, 2021, p. 102208.

2 Energy conversion losses for power-to-power, with remaining energy content in brackets: Electricity (100 per cent) to hydrogen through electrolysis (73 per cent), compression (67 per cent), conversion back to electricity through fuel

cell (29 per cent). IEA, *Technology Roadmap – Hydrogen and Fuel Cells*, Paris, June 2015.

3   Estimated shipping costs from Australia to Europe, in US dollars per kg of hydrogen delivered, from a recent study: liquid hydrogen $2.09/kgH$_2$, liquid organic hydrogen carriers $1.37/kgH$_2$, as liquified natural gas $1.07/kgH$_2$, as methanol $0.68/kgH$_2$, and as ammonia $0.56/kgH$_2$. C. Johnston et al., 'Shipping the sunshine: An open-source model for costing renewable hydrogen transport from Australia', *International Journal of Hydrogen Energy* 47(47), 1 June 2022.

4   Even green hydrogen is associated with some greenhouse gas emissions in its production chain, because of the greenhouse gas emissions that occur in the production of the equipment used for energy generation (wind turbines, solar panels and so forth) and the production of the hydrogen production equipment (electrolysers). However, these will usually be very small relative to the residual emissions from blue hydrogen, let alone hydrogen without CCS.

5   IEA, 2021; World Energy Outlook, 2021.

6   Data on emissions intensities is from T. Longden, et al., '"Clean" hydrogen? Comparing the emissions and costs of fossil fuel versus renewable electricity based hydrogen', *Applied Energy*, 306, p. 118145.

7   For example, Australia's national hydrogen strategy.

8   Longden, et al., '"Clean hydrogen?"'

9   See W. Cheng and S. Lee, 'How green are the national hydrogen strategies?' *Sustainability* 14(3), p. 1930.

10 Longden, et al., '"Clean" hydrogen?'

11 M. Venkataraman et al., 'Zero-carbon steel production: The opportunities and role for Australia', *Energy Policy* 163, 2022, p. 1 12811.

6 *Decarbonising China's Steel Industry*

1   BHP, *Pathways to Decarbonization*, Episode 3: 'Regional approaches to steel', 2021.

2   World Integrated Trade Solution (WITS), World Bank, Data on Export, Import, Tariff, NTM (worldbank.org)

3   R. An et al., 'Potential of energy savings and CO2 emission reduction in China's iron and steel industry', *Applied Energy* 226, 2001, pp. 162–80.

4   H. McKay, Y. Sheng & L. Song, 'China's metal intensity in comparative perspective' in Ross Garnaut, Jane Golley & Ligang Song (eds), *China: The Next Twenty Years of Reform and Development*, ANU ePress, 2010, pp. 73–98.

5   A. Arasto et al., 'Costs and potential of carbon capture and storage at an integrated steel mill' *Energy Procedia* 37, 2013, pp. 7117–24.

6   A. Bhaskar et al., 'Decarbonizing primary steel production: Techno-economic assessment of a hydrogen based green steel production plant in Norway', *Journal of Cleaner Production*, 2022, p. 131339.

7   X. Zhang et al., 'A review on low carbon emissions projects of steel industry in the world', *Journal of Cleaner Production*, 2021, p. 306.

7 *Land Carbon*

1   2008 figures. Down to 1 metre for Soil Organic Carbon (1550 GT). Inorganic depth unknown.

2   R. Lal et al., 'Soil carbon sequestration impacts on global climate change and food security', *Science* 304(1623), 2004.

3   Todd A. Ontl & Lisa A. Schulte, 'Soil carbon storage', *Nature Education Knowledge* 3(10), p. 35.

4   Climate Action Tracker, *Warming Projections Global Update*, May 2021.

5   J-F. Bastin et al., 'The global tree restoration potential', *Science* 365(6448), 2019.

6   B.W. Griscom et al., 'Natural climate solutions', *Proceedings of the National Academy of Science* 114(44), 16 October 2017, pp. 11645–50.

7   S. Roxburgh et al., 'Assessing the carbon sequestration potential of managed forests: A case study from temperate Australia', *Journal of Applied Ecology*, 43(6), 2006, pp. 1149–59.

8   C. Aponte et al., 'Structural diversity underpins carbon storage in Australian temperate forests', *Global Ecology and Biogeography* 29(5), 11 February 2020, pp. 789–802; John M. Dwyer, Rod Fensham & Yvonne M. Buckley, 'Restoration thinning accelerates structural development and carbon sequestration in an endangered Australian ecosystem', *Journal of Applied Ecology* 47(3), June 2010, pp. 681–91.

9   Heather Keith, Brendan J. Mackay & David B. Lindenmayer, *Re-evaluation of Forest Biomass Carbon Stocks and Lessons from the World's Most Carbon-dense Forests*, The Fenner School of Environment and Society, Australian National University, Canberra, 9 March 2009.

10 B.G. Mackey et al., *Green Carbon: The role of natural forests in carbon storage*, Australian National University, Canberra, 2008.

11 Forest cover using the 1992 National Forest Inventory definition: 'An area, incorporating all living and non-living components, that is dominated by trees having usually a single stem and a mature or potentially mature stand height exceeding 2 metres and with existing or potential crown cover of overstorey strata about equal to or greater than 20 per cent. This includes Australia's diverse native forests and plantations, regardless of age. It is also sufficiently broad to encompass areas of trees that are sometimes described as woodlands.' P. Woodgate & R. Keenan, 'Forest cover after European settlement', Victoria's

Forestry Heritage, 6 October 2021, www.victoriasforestryheritage.org.au/forest-estate/native-forests/forest-extent.html, accessed 8 July 2022.

12 Department of Agriculture, 'Australian forest profiles; Eucalypt', October 2019, www.awe.gov.au/sites/default/files/abares/forestsaustralia/publishingimages/forest%20profiles%202019/eucalypt/AusForProf_2019_Eucalypt_v.1.0.0.pdf, accessed 8 July 2022.

13 Allocation of 6 per cent subtropical (largest difference occurred in Queensland: https://soe.environment.gov.au/theme/land/topic/2016/vegetation-0), 1 per cent tropical and 3 per cent temperate. Using the above numbers for annual biomass each year. 126 million hectares of eucalypt woodland in 1750. 12.6 million for 10 per cent. 16.6 million tonnes of biomass (3.78 million ha) for temperate, 37.8 MT (7.56 million ha) for subtropical, 8.82 MT (1.26 million ha) for tropical. Total *0.5 (carbon) * 44/12 (CO_2) = 115.9 million tonnes.

14 Keith, Mackay & Lindenmayer, *Re-evaluation*.

15 Adam Morton & Anne Davies, 'Australia spends billions planting trees then wipes out carbon gains by bulldozing them', *Guardian Australia*, 17 October 2017; Peter Hannam and Adam Morton, 'Australia's climate data to UN questioned as study finds land clearing in Queensland underreported', *Guardian Australia*, 19 May 2022.

16 ABS, 'Physical account for land use, 2011 and 2016', Canberra, www.abs.gov.au/statistics/environment/environmental-management/national-land-account-experimental-estimates/2016#data-download, accessed 1 December 2021.

17 30 per cent obtained from comparison of the distribution of mallee in the below document and the land cover map seen above.

18 Australian National Botanic Gardens, 'Mallee plants – surviving harsh conditions', www.anbg.gov.au/gardens/education/programs/mallee.pdf, accessed 8 July 2022.

19 Australia State of the Environment, 'Vegetation: Land (2016)', SOE website, https://soe.environment.gov.au/theme/land/topic/2016/vegetation-0, accessed 8 July 2022.

20 5 tonnes of biomass average per hectare provided by Simon Dawkins at Oil Mallee org. Multiplication of 0.5 and 44/12 for CO_2e.

21 M. Adams et al., 'Legumes are different: Leaf nitrogen, photosynthesis, and water use efficiency', *Proceedings of the National Academy of Sciences*, 113(15), 2016, pp.4098–4103; J. Querejeta et al. 'Higher leaf nitrogen content is linked to tighter stomatal regulation of transpiration and more efficient water use across dryland trees. *New Phytologist*, 235(4), 2022. pp.1351–64; J. Doby et al., 'Aridity drives phylogenetic diversity and species richness patterns of nitrogen-fixing plants in North America', *Global Ecology and Biogeography* 31(8), 2022, pp. 1630–42.

22 B.R. Maslin & M.W. McDonald, *AcaciaSearch: Evaluation of Acacia as a woody crop option for southern Australia*, Report to the Joint Venture Agrofestry Program, Publication No 03/017, WA Department of Conservation and Land Management.

23 Anthony O'Grady & Patrick Mitchell, *Agroforestry: Realising the triple bottom line benefits of trees in the landscape*, CSIRO, 2018.

24 E. Luedeling et al., 'Forest restoration: Overlooked concerns', *Science*, 19 October 2019.

25 Luedeling et al., 'Forest restoration'.

26 R.M. Gifford et al., 'Australian land use, primary production of vegetation and carbon pools in relation to atmospheric carbon dioxide concentration' in R.M. Gifford & M.M. Barson (eds), *Australia's Renewable Resources: Sustainability and Global Change*, Bureau of Rural Resources Proceedings, No. 14, AGPS, Canberra, 1990.

27 P. Polgase et al., *Opportunities for Carbon Forestry in Australia: Economic assessment and constraints to implementation*, CSIRO, Canberra, 2011.

28 P. Polgase et al., *Regional Opportunities for Agroforestry Systems in Australia*, Rural Industries Research and Development Corporation, October 2008.

29 Bruno Glaser, 'Prehistorically modified soils of central Amazonia: a model for sustainable agriculture in the twenty-first century', *Philosophical Transactions of the Royal Society of London B Biological Sciences* 362(1478), 2007.

30 Cornell University, Department of Crop and Soil Sciences, 'Terra Preta de Indio', www.css.cornell.edu/faculty/lehmann/research/terra%20preta/terrapretamain.html, 8 July 2022.

31 Department of Primary Industries and Regional Development, 'What is soil organic carbon?' www.agric.wa.gov.au/measuring-and-assessing-soils/what-soil-organic-carbon, 28 June 2022, accessed 8 July 2022.

32 Jufeng Zheng et al., 'Biochar compound fertilizer increases nitrogen productivity and economic benefits but decreases carbon emission of maize production', *Agriculture, Ecosystems & Environment* 241, 2017, pp. 70–8; D. Zhang et al., 'Biochar helps enhance maize productivity and reduce greenhouse gas emissions under balanced fertilization in a rainfed low fertility inceptisol', *Chemosphere* 142, 2016, pp. 106–13; J. Lehmann et al., 'Biochar in climate change mitigation', *National Geoscience* 14, 2021, pp. 883–92; Hans-Peter Schmidt et al., 'Biochar in agriculture – A systematic review of 26 global meta-analyses', *GCB Bioenergy* 13(11), November 2021; G.M. Fernandez, Z. Durmec, P. Vercoe & S. Joseph, 'Fit-for-purpose biochar to improve efficiency in ruminants', *Meat and Livestock Australia*, Sydney, 2002.

33 R. Leng, T. Preston & S. Inthapanya, 'Biochar reduces enteric methane and improves growth and feed conversion in local "Yellow" cattle fed cassava root

chips and fresh cassava foliage', *Livestock Research for Rural Development* 24, 2012.

34 Pratikorn Sriphirom, 'Effects of biochar on methane emission, grain yield, and soil in rice cultivation in Thailand', *Carbon Management* 12(2), 2021; Qiong Nan et al., 'Exploring long-term effects of biochar on mitigating methane emissions from paddy soil: A review', *Biochar* 3, 2021, pp. 125–34; Thomas M. Winders et al., 'Evaluation of the effects of biochar on diet digestibility and methane production', *Translational Animal Science* 3(2), March 2019, pp. 775–83.

35 Ember, 'Carbon pricing', https://ember-climate.org/data/carbon-price-viewer/, 28 February 2022, accessed 8 July 2022.

36 IPCC, 'Strengthening and implementing the global response', *Special Report: Global Warming of 1.5C*, https://www.ipcc.ch/sr15/chapter/chapter-4/, accessed 8 July 2022.

37 European Commission, 'What was Horizon 2020?' https://ec.europa.eu/programmes/horizon2020/en/area/bio-based-industries, accessed 8 July 2022.

38 Queensland Department of Agriculture and Fisheries, 'Bio-based industrial products', DAF website, 20 July 2017, www.daf.qld.gov.au/business-priorities/agriculture/plants/bio-products, accessed 8 July 2022.

39 Institute for Bioplastics and Biocomposites, *Biopolymers facts and statistics 2020*, Hochschuhe Hannover, Hannover, edition 7, 2020.

40 C. Szernik & A.V. Bridgwater, 'Overview of applications of biomass fast pyrolysis oil', *Energy Fuels*, 18(2), 2004, pp. 590–2.

41 $17 + 13.3 = 30.3 / (169 + 116) = 0.106$. US Energy Information Administration, 'Petroleum and other liquids', www.eia.gov/international/data/country/KOR/petroleum-and-other-liquids/annual-refined-petroleum-products-consumption?pd=5&p=0000001001vg000000000000000000000000000000000g&u=1&f=A&v=mapbubble&a=-&i=none&vo=value&&t=C&g=none&l=249-000000000000000000000000g4&s=94694400000&e=1609459200000&ev=false, accessed 8 July 2022. 169 million metric tonnes EIA for Japan in 2019. 116 million metric tonnes EIA for South Korea in 2019.

42 81 billion gallons of jet fuel consumed in 2018. US Department of Energy, Sustainable Aviation Fuel: Review of technical pathways, Office of Energy Efficiency & Renewable Energy, September 2020, www.energy.gov/sites/prod/files/2020/09/f78/beto-sust-aviation-fuel-sep-2020.pdf, accessed 8 July 2022. = 245 million tonnes of jet fuel. $13.5/245 = 0.055 = 5.5$ per cent or 6 per cent.

43 $87.5gCO_2e/MJ$ of kerosene (de Jong, 2017). 142.2 MJ/gallon of kerosene. 12.4kg Co_2e/gallon of jet fuel.

Divide by 3.78541 = 3.2757 kg CO_2e/litre of jet fuel.

44 $12.4/1000 = 0.0124$ tonnes. Multiplied by 330.22 (number of gallons per tonne) = 4.094 tonnes CO_2e/tonne of kerosene. Multiply by 13.5 million tonnes and

by the emission reduction factor 0.68 of Alcohol-to-jet fuel conversion. 4.094 * 0.68 * 13.5 million = 37.6 million tonnes of CO_2e.

45 'Australia's annual carbon emissions reach record high', *Guardian Australia*, 19 March 2019; 'Annual CO_2 emissions worldwide from 1940 to 2020', Statista. com, https://www.statista.com/statistics/276629/global-co2-emissions/, accessed 8 July 2022.

46 Ralph Chami et al., *Nature's Solution to Climate Change*, IMF, December 2019.

47 Department of Climate Change, Energy, the Environment and Water, 'Coastal blue carbon ecosystems', www.awe.gov.au/science-research/climate-change/ ocean-sustainability/coastal-blue-carbon-ecosystems, accessed 8 July 2022.

48 Andy Steven & Mat Vanderklift, *Incorporating Blue Carbon Ecosystem Services into the Blue Economy and National Accounting*, Indian Ocean Blue Carbon Hub, CSIRO.

49 Peter I. Macreadie et al., 'Carbon sequestration by Australian tidal marshes', *Scientific Reports* 7, 44071, 2017.

50 Ross Hill et al., 'Can macroalgae contribute to blue carbon? An Australian perspective', *Limnology and Oceanography* 60(5), 25 June 2015, pp. 1689–1706.

51 www.stockholmresilience.org/research/planetary-boundaries/the-nine-planetary-boundaries.html, accessed 7 July 2022.

52 Vesna Gagic, Cate Paull & Nancy A Schellhorn, 'Ecosystem service of biological pest control in Australia: the role of non-crop habitats within landscapes', *Austral Entomology*, 2018.

53 *Green Ants as Biological Control Agents in Agroforestry, Rural Industries Research and Development Corporation*, Charles Darwin University Joint Venture Agroforestry Program, RIRDC Publication No. 09/1, 2009.

54 https://carbonpricingdashboard.worldbank.org/map_data, accessed 7 July 2022.

55 Clean Energy Regulator, *Quarterly Carbon Market Report*, September quarter 2021.

56 Sonali Paul, 'Australian carbon offset prices hit record high in polluters' spree', Reuters, 11 October 2021.

57 Andrew Macintosh, Don Butler & Dean Ansell, Measurement Error in the Emissions Reduction Fund's Human-induced Regeneration (HIR) Method, 14 March 2022; Andrew Macintosh et al., *The ERF's Human-induced Regeneration (HIR): What the Beare and Chambers report really found and a critique of its method*, Australian National University, 16 March 2022; Andrew Macintosh, *The Emission Reduction Fund's Landfill Gas Method: An assessment of its integrity*, Australian National University, 16 March 2022.

58 California Air Resources Board, *California's Compliance Offset Program*, 27 October 2021.

8 The Renaissance of Rural and Regional Australia

1   Barcaldine Post Office – Bureau of Meteorology station number: 36007.

2   John Ellicott, 'Options arise as world fertiliser prices continue to soar', *The Land*, January 2022, www.theland.com.au/story/7597114/fertiliser-outlook-looks-tight-again-but-there-are-other-options/, accessed 7 July 2022.

3   Luke Radford, 'Wet weather, rising fertiliser prices put squeeze on Tasmanian potato growers,' ABC Rural (online), 17 November 2021.

4   Unite & Recover Strategies, 'Attracting Manufacturers to Queensland', Queensland government, June 2021.

5   'State Infrastructure Strategy', Queensland government, June 2021.

ACKNOWLEDGEMENTS

Three years ago, *Superpower: Australia's Low-Carbon Opportunity* contributed to a change in the way many Australians view our country's contribution to humanity's defence against climate change. Analysis, discussion and experience over the past three years have greatly expanded our knowledge of the challenge and the shape and scale of the opportunity. This book brings together much of that knowledge.

The Superpower Transformation builds on the knowledge of people who have been developing highly practical ways of turning Australia's natural endowments for reducing carbon emissions and absorbing carbon from the atmosphere into success in the battle against climate change, and into Australian economic opportunity and prosperity. Five of the eight chapters have been written by people who have links through the University of Melbourne. Part of their work, and that of Isabelle Grant in Chapter 7, has been undertaken within Sunshot Zero Carbon Futures, supported by the zero-emissions electricity and hydrogen company ZEN Energy. We all learn much from each other. Chapters 5 and 6 are contributions from longstanding close colleagues at the Australian National University.

All the chapters have emerged from highly practical work involving many others. Chapters 1, 4, 7 and 8 have been informed by practical work within Zen Energy and its subsidiary Sunshot Industries, and consulting commissions with several businesses seeking paths to zero net emissions and with the governments of South Australia, Queensland, Western Australia and the Northern Territory. Chapter 8 is informed by our work with the CSIRO, Remote Areas Planning and Development Corporation and the Barcaldine Regional Council. We thank all these organisations, and in particular our colleagues and partners in central Queensland, including Sean Dillon, Rob Chandler, Morgan Gronold, Suzanne Thompson, David Phelps and Sharon Brown.

Peter Farley, president of the Vernier Society, assisted the authors of chapters 1, 3, 5 and 6 with his insights into zero-emissions manufacturing in Australia in the international context. We commend his own papers on green iron and steel, green manufacturing generally and international developments that are helping to frame the Superpower opportunity. Thank you, Peter, for your continuing large contributions to building an advanced manufacturing sector in Victoria – and Australia.

Authors of the individual chapters have their own debts to colleagues and partners.

In addition to contributing Chapter 8, Susannah Powell has worked closely with me in bringing together the elements of this book. Nadia Cusano has supported Susannah and me in much of the practical work of coordination. Thank you, Susannah and Nadia.

Nancy Viviani and Jayne Garnaut read the draft manuscript and contributed valuable comments.

Chris Feik at Black Inc. made his usual transformative contribution to presentation and intellectual clarity. Kirstie Innes-Will's skilled editing, under great pressure from reconciling deadlines with untimely arrival of contributions, lifted the quality of the book while managing the many demands of production. This is the fourth book that I have published with Black Inc, three through Black Inc's partnership with La Trobe University Press. The high quality of the editorial work and the congenial interactions through which these have been delivered have made the relationship a great pleasure, as well as being immensely productive.

Ross Garnaut
Melbourne, 31 July 2022

INDEX

EDITOR

Ross Garnaut is Emeritus Professor of Economics at the University of Melbourne and the Australian National University. In 2008, he produced the Garnaut Climate Change Review for the Commonwealth, state and territory heads of government. He is the author of the bestselling *Dog Days*, *Superpower* and *Reset*.

CONTRIBUTORS

Rebecca Burdon is the CEO at Climate Resource.

Isabelle Grant is a land carbon analyst at Sunshot Zero Carbon Futures.

Frank Jotzo is a professor of environmental economics and climate change economics at the ANU Crawford School of Public Policy, where he directs the Centre for Climate and Energy Policy.

Jared Lewis is an expert in developing solutions for processing and visualising large climate science datasets and the CTO of Climate Resource.

Dylan McConnell is an authoritative analyst of the Australian electricity markets.

Malte Meinshausen is Associate Professor at the University of Melbourne in Climate Science and a Lead Author of the IPCC Sixth Assessment Report.

Zebedee Nicholls is a leading expert in reduced complexity climate model development and a contributing author to the IPCC Sixth Assessment Report.

Susannah Powell is the general manager of Sunshot Zero Carbon Futures.

Mike Sandiford is a Redmond Barry Distinguished Professor Emeritus at the University of Melbourne, where he formerly held the positions of Chair of Geology and Inaugural Director of the Melbourne Energy Institute.

Ligang Song is a professor of economics at Arndt-Corden Department of Economics and director of the China Economy Program in the Crawford School of Public Policy, Australian National University.